T0252013

ARTIFICIAL INTELLIGENCE WITH UNCERTAINTY

SECOND EDITION

ARTIFICIAL INTELLIGENCE WITH UNCERTAINTY

SECOND EDITION

Deyi Li and Yi Du

Tsinghua University
Beijing, China

CRC Press
Taylor & Francis Group
Boca Raton London New York

CRC Press is an imprint of the
Taylor & Francis Group, an **informa** business

A CHAPMAN & HALL BOOK

國防工業出版社
National Defense Industry Press

MATLAB® is a trademark of The MathWorks, Inc. and is used with permission. The MathWorks does not warrant the accuracy of the text or exercises in this book. This book's use or discussion of MATLAB® software or related products does not constitute endorsement or sponsorship by The MathWorks of a particular pedagogical approach or particular use of the MATLAB® software.

CRC Press
Taylor & Francis Group
6000 Broken Sound Parkway NW, Suite 300
Boca Raton, FL 33487-2742

© 2017 by CRC Press, Taylor & Francis Group, 6000 Broken Sound Pkwy., NW, Suite 300, Boca Raton, FL 33487, under exclusive license granted by National Defense Industry Press for English language throughout the world except Mainland China.
CRC Press is an imprint of Taylor & Francis Group, an Informa business

No claim to original U.S. Government works

Printed on acid-free paper

International Standard Book Number-13: 978-1-4987-7626-4 (Hardback)

Library of Congress Cataloging-in-Publication Data

Names: Li, Deyi, 1944- author. | Du, Yi, 1971- author.
Title: Artificial intelligence with uncertainty / Deyi Li and Yi Du.
Description: Boca Raton : CRC Press, [2017]
Identifiers: LCCN 2016050022| ISBN 9781498776264 (hardback) | ISBN 9781498776271 (ebook)
Subjects: LCSH: Artificial intelligence. | Uncertainty (Information theory)
Classification: LCC Q335 .L5 2017 | DDC 006.3--dc23
LC record available at https://lccn.loc.gov/2016050022

Visit the Taylor & Francis Web site at
http://www.taylorandfrancis.com

and the CRC Press Web site at
http://www.crcpress.com

Contents

Preface to the Second Edition

Eleven years have elapsed since the first edition of *Artificial Intelligence with Uncertainty* came out. Reading the words written back then, I still feel warm hearted. The second edition is the result of the deepening of basic theoretical research on artificial intelligence with uncertainty, a coming together of cloud computing, big data, networking, and smart planet in the context of an Internet environment, creating a group intelligence through sharing and interaction in social computing. All these point to the fact that a new era has arrived after a half century of developments in which Turing computing has become Internet computing, and in which an artificial intelligence that simulated human thought processes through mathematical and symbolic logic became artificial intelligence with uncertainty.

Language is both the carrier of human thought and the accumulation of human intelligence. We believe that natural language should be the starting point for research into artificial intelligence with uncertainty. The qualitative concepts expressed by language values serve as the basis of human thought, which is random and vague. We explain the probability and statistics of fuzzy sets and type-2 fuzzy sets indicated by membership degree. When it comes to the dependence of fuzziness on randomness, we find that the subjection degree is neither precise nor subjective, but that it is uncertain and objective, which is elaborated on in this book.

A cloud model, especially a Gaussian cloud model, reflects the uncertainty of concepts in the process of human cognition through the three digital characteristics of expectation, entropy, and hyper entropy. A Gaussian cloud, composed of a large number of cloud droplets, comes from, but is different than, a Gaussian distribution. A series of interesting and meaningful statistical properties can be obtained through the "devil" of excess entropy. As for the high-order cloud model formed by increasing the order, a cloud droplets group can even transform from Gaussian distribution to power-law distribution in different orders, which this book explains quite clearly.

Physics can change the world. Atomic models and fields are physical perceptions of the objective world, which we introduced into a subjective perception of the world. The human brain is often called a small universe, which means that it is as rich and vast as the universe, a map of nature and society. Thanks to the support of cloud models, cloud transformation, cloud reasoning, cloud control, data fields, and topological potential, we come up with a physical method for cognitive computing. Through controlling a triple inverted pendulum, we present a variety of balancing patterns, explain the emergence of a spontaneous applause mechanism in concert halls, develop a wheeled robot to simulate different human driving behaviors, and develop a variable sized personalized mining engine in a cloud computing environment. These vivid cases give us a more profound understanding of artificial intelligence.

This book has been used as teaching material by more than 300 graduate students over six academic years at several universities—including Tsinghua University, Beijing University of Aeronautics and Astronautics, Wuhan University, Beijing University of Posts and Telecommunications, and PLA University—supported by course materials, a teaching website, and a teaching practice database. All the insightful queries and brainstorming that came out of the teaching process will not be forgotten.

Science is rooted in the discussion. Exchanging ideas is important for both authors and readers, in teaching and learning. We hope that more people will join in the exploration of artificial intelligence with uncertainty and experience its charm.

Deyi Li and Yi Du

Postscript to the Second Edition

It is gratifying that although *Artificial Intelligence with Uncertainty* has been in print for nine years, it still receives warm attention from readers. However, the authors have regretted that some of the content is still not accurate enough. With the rise of soft computing, semantic computing, granular computing, cloud computing, the Internet of Things, and big data over the years, the authors have had an urge to republish the book now and then. It was not until 2015 that they made up their minds to revise the book after carefully organizing their thoughts: replying to questions raised by readers and students; recalling discussions with, and doubts of, academic peers; and summarizing the latest work.

We have spared no effort to publish a book that will not only amuse readers, but will also enable those who are interested in duplicating the models, algorithms, and cases mentioned. We have also tried our best to make it an enjoyable read by making it interesting, vivid, and lively rather than creating a deductive encyclopedia, however abstruse the subject seems to be. To avoid it being too heavy, we have highlighted the original nature of our work, deleted some general mathematical and basic knowledge, and emphasized intelligent computing related to uncertainty, such as cloud computing, cloud transformation, big data, and social computing.

The revision of the book owes much to the arduous labor of the first author's postgraduate students, including Gan Wenyan and Liu Yuchao, who contributed to the preparation of some sections. The authors have benefited greatly from discussions with well-known professors, including Professors Kleinberg, Barabási, Mendel, Kai Huang, Lixin Wang, and others, and some young scholars, including Yu Liu, Changyu Liu, Kun Qin, Liwei Huang, Mu Guo, Wen He, Shuliang Wang, Liwei Huang, Xin Wang, and many more. Zengchang Qin, Gansen Zhao, Dongge Li, Yu Zhou, Yanwei Fu, and Yiran Chen revised the contents of this book. On the eve of republishing, hearty gratitude is extended to those helpful students and academic peers!

We are also grateful to our relatives. Thanks to their unremitting support and loving care we were able to finish this book successfully.

Sincere thanks are given to all who offered to help with the book!

Introduction

This book is concerned with uncertainty in artificial intelligence. Uncertainty in human intelligence and human knowledge is universal, objective, and beautiful but also difficult to simulate. The models and reasoning mechanisms for artificial intelligence with uncertainty are discussed in this book through cloud model, cloud transformation, data field, intelligent control, and swarm intelligence.

The authors' research have given this book a distinct character, making it valuable for scientists and engineers engaged in cognitive science, AI theory, and knowledge engineering. It is a useful reference book for university graduate students.

MATLAB® is a registered trademark of The MathWorks, Inc. For product information, please contact:

The MathWorks, Inc.
3 Apple Hill Drive
Natick, MA, 01760-2098 USA
Tel: 508-647-7000
Fax: 508-647-7001
E-mail: info@mathworks.com
Web: www.mathworks.com

Preface to the First Edition

It is often said that three big difficult problems remain unsolved in the world: generation of life, origin of the universe, and how the human brain works. This book addresses the third problem, and it studies and explores the uncertainties of knowledge and intelligence during the cognition process of human brain and how to use the computer to simulate and deal with such uncertainties.

Why does humankind possess intelligence? How does the human brain work? As the quintessence of hundreds of millions of years of biological evolution and millions of years of human evolution, how does the human brain process uncertainties? We don't know much about these very important and interesting questions.

If brain science explores the secrets of the brain via research on cells and molecules, if cognitive psychologists master the rule of the brain via observation on "stimulus-response" (i.e., "input–output"), then artificial intelligence, which has been researched on for nearly 50 years, tends to formalize representation of knowledge and use symbolic logic more often to simulate the thinking activities of human brain.

In the twenty-first century, we step into the Information Age, when the information industry leads the development of the global economy, and rapid development of information technology is changing the society we live in, including our work mode and lifestyle. Some say that an information industry and information technology–oriented knowledge economy era is announcing its arrival to the whole world. However, while we are enjoying the Internet technology and culture, we are also confronted with information overload; therefore, people are trying to find real information, the information they need, or even new knowledge from the ocean of data through artificial intelligence. Here, a most basic aspect of artificial intelligence is involved, that is, representation of knowledge. John von Neumann, the most influential scientist of the twentieth century, who made outstanding contributions to the development of electronic computers and was renowned as the "Father of the Electronic Computer," carried out in-depth studies on similarities and differences between electronic computers and the human brain, and he asserted in his posthumous work that language of the human brain is not a mathematical language.

Natural language is the basic tool of human mind. We think a very important entry point of research on artificial intelligence should be natural language, which is the carrier of knowledge and intelligence. The qualitative concept indicated by natural language value possesses uncertainty, especially randomness, fuzziness, and the relatedness between randomness and fuzziness. Exploring more deeply from such an entry point is exactly the artificial intelligence with uncertainty we are going to discuss.

This book discusses the objectiveness, universality, and positivity of the existence of uncertainty in human knowledge and intelligence; studies the mathematical basis, feature, representation, model, inference mechanism, certainty of the uncertain thinking activities of artificial intelligence with uncertainty; expands hierarchically from the cloud model used for qualitative and quantitative transformation and the physical method of cognition to data mining, knowledge discovery, and intelligent control; identifies the regularity in uncertain knowledge and intelligent processing; and finally forecasts the development direction of research on artificial intelligence with uncertainty.

In the past 10 years, supported by the National Natural Science Foundation of China, 973 Program, 863 Program, and the National Defense Pre-Research Foundation of China, we have completed some research on artificial intelligence with uncertainty, which seems to be reaching an organic whole, unifying many important but partial results into a satisfactory framework to explain how artificial intelligence with uncertainty explored and generalized the traditional artificial intelligence science.

Humans are born with the desire to seek knowledge and seek beauty. How do people process cognition? How do people try to use the theory of "beauty" to explain and simulate the cognition of humankind? This is exactly where our interests lie when we strive to make the exploration. The research subject of this book is very challenging and interesting, but the knowledge level and practical ability of the authors are limited, so the writing of this book is just an exploration and is the deepening of the research work. Inevitably, there are errors in the book, and we welcome readers to comment and point out any mistakes.

This book will be useful to scholars engaging in the study of cognitive science, brain science, artificial intelligence, computer science, and cybernetics, especially researchers and developers working on understanding and processing of natural language, intelligent retrieval, knowledge engineering, data mining, knowledge discovery, and intelligent control. Meanwhile, it could also serve as a teaching book or reference book for postgraduates of related majors in universities and colleges. It is hoped that there will be more people participating in the exploration of artificial intelligence with uncertainty. May our innermost sincerity and blessing be sent to you—for people who read this book by chance at a certain moment, it might be beneficial to your job and interest, which is exactly the glamour of knowledge and intelligence with uncertainty.

Deyi Li and Yi Du

Authors

Deyi Li earned his PhD in computer science at Heriot-Watt University in Edinburgh, U.K., in 1983. He was elected a member of the Chinese Academy of Engineering in 1999 and a member of the Eurasian Academy of Sciences in 2004. Dr. Li has published more than 100 papers on a wide range of topics in artificial intelligence. His other books include *A Prolog Database System* and *A Fuzzy Prolog Database System*.

Yi Du earned his PhD in computer science at PLA University of Science and Technology in Nanjing, China, in 2000. At present, Dr. Du is a senior engineer in network management in Beijing.

Research Foundation Support

The research projects in this book were supported by the following.

National Natural Science Foundation of China

- Research of Knowledge Discovery in Databases, 1993–1995, No. 69272031
- Qualitative Control Mechanism and Implementation of Triple Link Inverted Pendulum Systems, 1998–2000, No. 69775016
- The Methodology Research of Knowledge Representation and Knowledge Discovery in Data Mining, 2000–2002, No. 69975024
- Study on Data Mining Foundations, 2004–2006, No. 60375016
- Recognition Theory of Irregular Knowledge, 2004–2008, No. 60496323
- Research on Network Data Mining Methods, 2007–2009, No. 60675032
- Research on the Evolution of Complex Network Structures Based on Topological Potential, 2010–2012, No. 60974086
- Research on Intelligent Testing Standards and Environmental Design of Unmanned Vehicles, 2010–2013, No. 90920305
- Research on Key Technologies for Massive Data Mining Based on Cloud Computing, 2011–2014, No. 61035004
- Research on the Key Technology of Multi Vehicle Interactive Collaborative Driving Based on Visual Auditory Information, 2012–2015, No. 91120306
- Domain-Independent Data Cleaning Based on Ant Colony Algorithm and Cloud Model, 2014–2017, No. 61371196

National Basic Research Program of China (973 Program)

- Theory and Methodology Research of Data Mining and Knowledge Discovery, 1999–2003, No. G1998030508-4
- Common Foundation of Advanced Design for Large Scale Application Software Systems: Theory and Methodology Research of Intelligent Design, 2004–2009, No. 2004CB719401
- The Basic Theory and Key Technology of Unstructured Information Processing Based on Visual Cognition, 2007–2011, No. 2007CB311000
- Requirement Engineering—Basic Research on Software Engineering of Complex Systems, 2007–2011, No. 2007CB310800

Related National Patents in This Book

- Membership Cloud Generation Method, Membership Cloud Generator and Membership Cloud Control Device. No. ZL 95 1 03696.3
- Method for Managing Water Print Relational Database. No. ZL 2003 1 0122488.7
- Image Segmentation Method Based on Data Field. No. ZL 2008 1 0172235.3
- Image Segmentation Method Based on Cloud Model. No. ZL 2008 1 0172236.8
- Community Division Method in Complex Network. No. ZL 2008 1 0224175.5
- Method for Searching Web Services According to Non-functional Requirements of User. No. ZL 2011 1 0103757.X
- An Automatic Clustering Method Based on Data Field. No. ZL 2011 1 04487.2
- Automatic Clustering Method Based on Data Field Grid Division. No. ZL 2011 1 0114544.7

Chapter 1

Artificial Intelligence Challenged by Uncertainty

During the rapid development of science and technology in the nineteenth-century Industrial Revolution, machines were employed to reduce manual labor or replace it altogether. In the twentieth century, advances in information technology, especially the advent of computers, produced machines that reduced or substituted for the work of the human brain, prompting the birth and rapid rise of artificial intelligence. In the twenty-first century, the widespread applications of the Internet and cloud computing, intelligent computing, sharing, interaction, and swarm intelligence have revolutionized people's lives, indicating the arrival of the new era of artificial intelligence.

Intelligence can be defined as wisdom and ability; what is called artificial intelligence (AI) is a variety of human intelligent behaviors, such as perception, memory, emotion, judgment, reasoning, proof, recognition, understanding, communication, design, thinking, learning, forgetting, creating, and so on, which can be realized artificially by machine, system, or network.

Over the past 60 years, there has been great progress in AI. However, machine intelligence, built on the basis of certainty or accuracy, is severely limited by its formal axiom system, the precision of which is unable to simulate the uncertainty of human thought processes.

Thinking is the human brain's reflection on and recognition of the essential attributes and the inner linkages of objective things. Language is the carrier and expression of human thought. Human intelligence differs from machine intelligence and even other forms of biological intelligence, in that only human beings are able to carry forward, through oral and written languages, knowledge accumulated over thousands of years. The uncertainty of intelligence is inevitably reflected in language, knowledge, thought processes, and results. Therefore, this book focuses on the study of uncertainty: its representation, reasoning, and simulation of human intelligence; on basic certainty in human intelligence; and how uncertainty challenges artificial intelligence.

1.1 The Uncertainty of Human Intelligence

1.1.1 The Charm of Uncertainty

In the nineteenth century, the deterministic science represented by Newton's theory created a precise way to describe the world. The whole universe, as a deterministic dynamic system, moves according to definite, harmonious, and orderly rules that can determine future events. From Newton to Laplace, and then to Einstein, they depict a scientific world that is completely determined. Followers of determinism believe that uncertainty is only due to people's ignorance of, not to the original look of things. If we are certain of the initial condition of things, we can completely control their development.

For a long time the deterministic view of science limited how people understood the universe. Despite living in a real world full of complex and chaotic phenomena, scientists only saw the predictable mechanical world with its structure and operational rules. They considered uncertainty insignificant, excluding it from the scope of modern scientific research.

According to the deterministic view, the present state and development of the universe were decided in its early chaotic stage. Things as large as changes in the world's landscapes, or as small as personal ups and downs, were determined 10 billion years ago. Nowadays people disagree with this point of view because the reality of uncertainty is all around us. However rich one's knowledge and experience, it is impossible to predict what will happen in life, for example, who will win an Olympic medal or how likely one is to win the lottery.

With developments in science, deterministic thinking has come up against insurmountable difficulties in an increasing number of research areas, such as the study of molecular thermal motion. With the multiple degrees of freedom that a large number of factors introduced, single-value deterministic theory is not only unable to solve the inherent complexity of the system, but the complexity of the system will also result in fundamental changes in things. Even though the full trajectory of particles and the interactions between them can be precisely determined, it is hard to know the exact behaviors forming the whole. Thus, in the late nineteenth century, Boltzmann, Gibbs, and others introduced the idea of randomness into physics and established statistical mechanics. Statistical mechanics states that, for a group of things, Newton's laws can only describe a general rule; individual members of the group cannot be described definitively, but only in terms of their behavior, that is, their "probability."

The emergence of quantum mechanics had a further impact on deterministic theory by showing that uncertainty is an essential attribute of nature. Quantum mechanics studies the movement of groups of atoms or particles. Because the particles are very small, observation, however it is conducted, interferes materially with the object it is observing. We cannot accurately determine the coordinates and momentum of the particles at a given moment. The more certain the coordinates of the particles, the more uncertain their movement, hence the "uncertainty principle" proposed by the German physicist Werner Heisenberg. The uncertainty of the objective world is not a transitional state caused by our incomplete knowledge, but is an objective reflection of the essential characteristics of nature.

Scientists have also admitted that although what we call scientific knowledge today is a collection of statements with different degrees of certainty, all our conclusions about science contain a degree of uncertainty. In the twentieth century, the philosopher Karl Popper described physical deterministic theory in his book *Objective Knowledge: An Evolutionary Approach* [1] thus: "I believe Peirce was right in holding that all clocks are clouds to some considerable degree—even the most precise of clocks. This, I think, is the most important inversion of the mistaken determinist view that all clouds are clocks." Cloud symbolizes the chaos, disorder, and unpredictability of uncertainties, such as climate change, ecosystems, celestial movements, Internet computing, human mental activity, and so on. Whereas the clock is accurate, orderly, and highly predictable. The Cartesians believe that all our actions merely seek to convert the cloud into a clock. Those who believe in nondeterminism think that in the physical world not all events are determined accurately in advance, in every detail. Many scientists and philosophers agree with the view that uncertainty is the objective state of human cognitive ability.

Clouds formed by drops of moisture in the sky take on form if viewed from afar but are shapeless when viewed close to. Their form is elegant and changeable, showing different poses, sometimes like blossoming cottons, sometimes rushing down like water, light or dark, curled or stretched and carefree. They float in the air, coming together and parting, changing all the time, stimulating poetic imagination, hence the use of the term "cloud computing" to describe the uncertainties of the Internet.

The undulating contours of hills, winding coastlines, and rivers extending in all directions are not smooth but uncertain irregular shapes whose size, length, and area change according to the scale on which they are being measured. Once the scale is determined, the measured value can be determined; within a certain range, there is a power function relationship between the scale and the measured values. The length of a coastline can only be determined by the device used to measure it. This is the uncertainty of measurement.

A line group is a point when viewed from a distance, a three-dimensional sphere when looked at closer; it becomes a curve when you approach the surface, and a one-dimensional element when looked at more closely; a careful look shows a three-dimensional cylinder then a two-dimensional plane when looked at even more closely. If you examine a fiber of wool it is one-dimensional, which becomes a three-dimensional cylinder if you look closer. If you look at the Earth from outside the galaxy it is a point; seen from within the solar system the Earth's orbit is elliptical; viewed on a plane the Earth is a two-dimensional surface. If you stand in the desert what you see will be completely different to what you see when standing on a small hill. This is the uncertainty of system dimension.

It is difficult to accurately represent the uncertainties of relationships between variables using analytic functions. Examples where x_i is a controllable variable with a specifiable value within a particular range, and y_i is a random variable with probability distribution include agricultural production and fertilizer (y_1, x_1); blood pressure and age (y_2, x_2); and strength and fiber length (y_3, x_3). There are many more variables or random variables that are not entirely independent of, dependent on, or associated with each other. This means the relationship between variables is uncertain.

You might inherit your grandfather's nose and your grandmother's earlobe; your sister may have eyes like her uncle; both of you might have the same forehead as your father. This is genetic uncertainty.

Particle collisions lead to the random orbital motion of particles. The course of this motion has a statistical self-similarity, that is, amplification of some small part of it may have the same probability distribution as that of some larger part. This is the uncertainty of movement.

The objective world is uncertain. The objective world mapped in the human brain is a subjective world, which should also contain uncertainty. Human cognitive processes therefore inevitably involve uncertainty.

Existence and thinking have an identity. When thinking abstraction, imagination, and randomness are hugely important because they lead beyond more existence to intuition, association, analogy, and sudden enlightenment in human intelligence, uncertainty in divergence and convergence, and creation. Creation has no definite pattern.

Speech and written language are two of the four major milestones in human civilization. Spoken language symbolizes an external object and written language is a code; both convey information. Language, especially in its written form, is passed on through the generations. The uncertainty of the objective world creates the uncertainty of natural language. The limitations of individual cognition make people describe the objective world in uncertain terms, which will be reflected in their language. The differences in people's cognitive abilities are also reflected in their different descriptions of the same event. Uncertainty, which is an essential feature of human cognition, is thus naturally reflected in language.

There is no fixed structure for the organization of language, any more than there is a fixed order for the expression of thoughts. Language consists of sentences.

A sentence can have different meanings in different contexts and situation. The basic elements in a sentence are linguistic values corresponding to different concepts. Generally, nouns are used to express concepts and the predicate expresses the relationship between concepts. Concepts and predicates have an uncertainty. Uncertainty in language does not impede our understanding; rather, it allows for the infinite possibilities of imagination. The Tang Dynasty poet Wang Bo paints a beautiful picture with the lines: "The proud eagle flies high in the sky, with rosy clouds floating alongside; the river runs away off to the horizon, reflecting the same tint in the blue sky." Cao Xueqin's "her two eyebrows, which seemed to knit and yet not to knit, and opening wide those eyes, which seemed to stare and yet not to stare," depicts Lin Daiyu's sick state and her aesthetic look. In these examples there can be no numerical substitutions.

Common sense is the abstraction of common knowledge, which includes many uncertainties, but it is also knowledge that is obvious. For example, people distinguish cups, plates, and bowls according to their size and use: a cup can be filled with water, a plate with rice, a bowl with soup; a cup has a handle but a plate does not; a bowl has an edge. But in terms of how we understand cups, plates, and bowls, there is no absolute ratio between width and height.

Common sense is different from professional knowledge because of its universality and directness. It may lack the depth and systematic nature of professional knowledge. It is largely relative, changing in accordance with time, place, and people. This is the uncertainty generated by relativity.

There was once a common understanding in AI that the essential distinction between human and machine was the possession or lack of common sense. Whether or not AI can eventually be achieved depends on whether human common sense can be realized. The uncertainties of cognition arise from the uncertainties of the objective world. Phenomena described as certain or regular will only occur under particular conditions, and will exist only partially or for a short time. Although many researchers in the natural sciences of physics, mathematics, and biology, and the social sciences of philosophy, economics, society, psychology, and cognition are involved in the study of certainty, hardly anyone challenges the essential uncertainty of the world. An increasing number of scientists believe that uncertainty is an essential element of the world, and that the only certainty is actually uncertainty!

Against this background, the science of chaos, the science of complexity, and AI with uncertainty have undergone rapid development, with the uncertainties inherent in common sense becoming a fruitful area for AI research.

Acceptance of the existence of uncertainty does not mean that we have to stop looking for basic, deterministic solutions, but we must remember not to draw premature conclusions when we have not fully understood the facts. Speculating arbitrarily on certainty will prevent us from finding effective solutions. Moreover, faced with complex situations, accepting the existence of uncertainty will help us keep an open mind, maintain a positive attitude, stimulate creative thinking, and promote scientific progress.

Cognitive uncertainty inevitably leads to research on artificial intelligence with uncertainty. A priority for AI scientists is how to represent and simulate the uncertainty of human understanding of the world, to enable the intelligence of machines, systems, or networks [2]. In this book, we start with the representation of uncertain knowledge, and propose a cognitive model–cloud model for qualitative–quantitative dual transformation, establish physical methods for the discovery of uncertainty knowledge and make use of cloud computing, natural language understanding, image recognition, data mining, intelligent control, social computing, and so on.

1.1.2 *The World of Entropy*

The first law of thermodynamics states that any thermodynamic process in nature must comply with the law of energy conversion and conservation. The modern post-industrial age—with the rapid development of science and technology, population growth, the increasing scarcity of water resources, worsening desertification, and excessive consumption of a variety of nonrenewable energy resources—gives us cause to think seriously about the future of mankind, of nature, and about fundamental questions of sustainable development. There always seems to be an invisible hand functioning, which might be entropy at work. Entropy is at the heart of the second law of thermodynamics, which dates back to the nineteenth century. In 1854, the German physicist R.J. Clausius first proposed the concept of entropy, ΔS, when studying thermodynamics, which is defined as

$$\Delta S = \frac{\Delta Q}{T}$$

where
  T is temperature
  Q is heat

In isolated systems, thermal molecular motion always goes from a centralized and orderly arrangement to a dispersed, chaotic state. Entropy represents the evenness of energy distribution in space and increases with the evenness of energy distribution. In the spontaneous transformation of a system from order to disorder, entropy always increases, which is represented by $\Delta S \geq 0$. When entropy reaches its maximum value in a system, the system enters a state of energy balance. At this moment, no further work is done by free energy within the system; the whole system is in a state of equilibrium. According to the second law of thermodynamics, without consumption or other change in a force, it is impossible for heat to flow from a low to a higher temperature.

This law states that mechanical energy is transformed into internal energy due to friction, viscosity, and other factors, and that the transformation is irreversible.

Heat is always transferred from an object with a higher temperature to one with a lower temperature, but not in reverse. The free expansion and diffusion of gas is also irreversible. For example, if you put a glass of hot water in the room and regard the room and the glass as an isolated system, the heat of the water will be distributed until its temperature is the same as that of the room, then the process of heat exchange stops. The heat spread into the air cannot automatically regather in the glass to boil the water again. In nature, towering majestic mountains become gravel rocks after many years of weathering, graceful flowers wither and are buried in dust. Entropy plays a role in this process.

In 1877, Boltzmann proposed an equation for entropy:

$$S = k \ln W$$

where

k is Boltzmann's constant

W is the number of microstates corresponding to a macroscopic state, that is, the thermodynamic probability of the macroscopic state

Entropy is proportional to the logarithm of the thermodynamic probability. The greater the thermodynamic probability, the more chaotic the system is. Therefore, thermodynamic entropy is a measure of the disorderliness of a system.

In 1948, Claude E. Shannon introduced the concept of information entropy, that is, the definition of probability entropy.

Let X be a system composed of n events $\{X_i\}$ ($i = 1, 2, \ldots, n$). Consider the event X_i with probability $p(X_i)$, then the information entropy (probability entropy) is defined as

$$H(X) = -\sum_{i=1}^{n} p(X_i) \log p(X_i)$$

Information entropy describes the average uncertainty of events occurring in the event set X. The higher the entropy, the greater is the degree of uncertainty. In the field of information, entropy is a measurement of the uncertainty of the system state. Here, "state" is not limited to thermodynamic systems but it extends the definition of entropy. In the formula for calculating entropy, if we use 2, 10, or e (i.e., 2.71828…) as a base, the units of information entropy are called bit, hart, and nat, respectively. When the probabilities of events in X are equal, information entropy reaches its maximum value. Shannon's contribution was to show that bits become the basic unit in the process of information transmission, referring to the uncertainty of the communication process.

When the basic concept of entropy was introduced into thermodynamics, physics was expanded beyond Clausius' proposition. By the 1990s the definition of entropy had been further generalized with thermo-entropy, electronic entropy, mobile entropy, vibration entropy, and self-spin entropy in statistical and quantum physics; in other areas there was topological entropy, geographic entropy, meteorological entropy, black-hole entropy, social entropy, economic entropy, anatomical entropy, spiritual entropy, cultural entropy, and so on.

Einstein defined entropy as the essential principle of all sciences. Originally used for the measurement of evenness of distribution, it then became a measure of the level of disorderliness, and further of uncertainty. It is therefore inevitable that it should be introduced into the field of artificial intelligence. As an important measure of uncertainty, entropy is a key parameter in the study of artificial intelligence with uncertainty. This book introduces entropy into the study of the cognitive model–cloud model, for qualitative–quantitative dual transformation and further proposes the concept of super-entropy.

1.2 Sixty Years of Artificial Intelligence Development

Since ancient times, humans have dreamed of artificially based intelligence, with relevant folklore dating as far back as 200 BCE. In the ancient Egyptian city of Alexandria, Heron probably invented the earliest intelligent machine in a number of automatic machines to ease manual labor, such as worshippers in the temple automatically obtaining the required amount of holy water by inserting coins into a slit; a gate opening automatically when the sacrificial fire was lit; and two bronze statues of priests standing on the sacrificial altar raising pots to pour holy water into the fire.

1.2.1 The Dartmouth Symposium

1.2.1.1 Collision between Different Disciplines

In June 1956 at Dartmouth, New Hampshire, four young scholars, John McCarthy, Marvin Minsky, Nathaniel Rochester, and Claude Shannon, organized a 2-month Dartmouth Summer Symposium on simulating human intelligence with machines. They invited 10 scholars from various fields, including mathematics, neurophysiology, psychiatry, psychology, information theory, and computer science. Each discussed their own research focus from the perspective of their individual discipline, which led to intense intellectual clashes.

The highlights of the Dartmouth Symposium were as follows: Marvin Minsky's Snarc, the stochastic neural analog reinforcement calculator; Alpha-Beta searching method proposed by John McCarthy; and the Logic Theorist, presented by Herbert Simon and Allen Newell, which proved a mathematical theorem [3].

While working on the cognitive process of proving a mathematical theorem, Herbert Simon and his colleagues found a common law. First, the whole problem is broken into several subproblems, then these subproblems are solved using substitution and replacement methods in accordance with the axioms and proven theorems. Based on this, they established the heuristic search technique for machines to prove mathematical theorems. Using the Logic Theorist program they proved a number of theorems from Chapter 2 of Russell and Whitehead's *Principia Mathematica*. Their work was regarded as a major breakthrough in the computer simulation of human intelligence [4].

Although the scientists present at the symposium had different perspectives, they all focused on the representative form and cognitive laws governing human intelligence. Making full use of the mathematical logic and computing capabilities of computers, they provided theories on formalized computation and processing; simulated some basic human intelligence behaviors; and created artificial systems with some sort of intelligence, enabling computers to complete work that could previously only be accomplished by human intelligence.

The Dartmouth Symposium marked the birth of AI, with John McCarthy (known as the Father of AI) [3] proposing the term "artificial intelligence" as the name of the new cross-discipline.

1.2.1.2 Ups and Downs in Development

In today's world intelligent machines, intelligent communities, intelligent networks, and intelligent buildings are everywhere and AI-related research is going on in almost every university. The development of AI has been far from smooth, however, with many controversies and misunderstandings along the way.

In the early 1980s, mathematical logic and symbolic reasoning became the mainstream of AI, and programming languages such as LISP and PROLOG were favored throughout the world. A Japanese initiative, called Fifth-Generation Computer Systems (i.e., computer systems processing knowledge and information), pushed AI research to a new high. However, the much anticipated "Fifth-Generation Computer" failed to go beyond the structural framework of the Von Neumann system and still used a program and data. It could not therefore realize a human–machine interaction through image, sound, and language in a natural way, nor could it simulate human cognitive processes.

Research on neural networks (i.e., NNs) had also boosted AI development. In the early 1940s, an artificial neural network (i.e., ANN) method was proposed by some scholars. In 1982, John J. Hopfield suggested using hardware to realize ANN [5]. In 1986, David E. Rumelhart and others put forward the Back Propagation algorithm in multilayered networks, which became a milestone in NN development. At that time, people expected that "thinking in images" could be processed through a biological, artificial neuron model that would simulate human visual and listening senses, and understanding, instinct, and common sense. Over the following two decades there was great enthusiasm among scholars for research on artificial neural networks, but the achievements fell short of expectations, both in theory and in applications, until deep belief network was proposed by Geoffrey Hinton in 2006. Since then a growing number of researchers, led by Geoffrey Hinton, Yoshua Bengio, and Yann LeCun, have engaged in related work on deep learning in a second wave of development on neural networks. Due to its advantages for discovering intricate structures in large data sets, deep learning has brought about breakthroughs in speech recognition, image analysis, text mining, and many other domains, greatly promoting the development of artificial intelligence.

In the 1970s AI, space technology, and energy technology were regarded as the top three technologies in the world. In the new century, with the extensive penetration of information technology into the economy, society, and everyday life, there is an increasingly urgent demand for machines that can simulate human intelligence. People began to question the Von Neumann architecture and seek new structures, like the Internet and quantum computers. AI was considered in a broader context. Cloud computing is widely used today, with the "Internet of people" and "Internet of things" becoming part of the infrastructure of human life and production. Swarm intelligence led by sharing and interaction have moved AI to a new stage.

1.2.2 Goals Evolve over Time

Over the 60 years of the development and growth of AI, the academic community has held various views on the goals for AI research.

1.2.2.1 Turing Test

The 100th anniversary of the birth of British mathematician, Alan Turing was marked in 2012. In 1950, he had proposed a test to determine whether a machine had human intelligence, which became a widely acknowledged AI test standard known as the Turing Test. The standard states that if the actions, reactions, and interactions of a machine are the same as those of a human being with consciousness, then it should be regarded as having consciousness and intelligence. To eliminate bias in the test Turing designed an imitation method. A human interrogator, separate from the machine, asks various questions of both a human and the machine. In a given period of time, the interrogator judges which answers came from the human and which from the machine. A series of such tests were devised to test a machine's intelligence level.

Some scholars argued that such machine intelligence was still far behind a human being. Let us imagine a machine engaged in a conversation with a human. However smart it is, the machine cannot have the same common sense as a person, nor can it consistently apply correct and special pronunciation, intonation, and emotion. It cannot react to different situations arising in the conversation process according to context. In this sense, the machine is not exhibiting human intelligence, even that of a child.

It has been argued that there are too many uncertainties in the description of the test standard. Parameters cannot be well defined or fixed, nor can the test standard be precisely determined.

Real machine intelligence should be available everywhere and exist in harmony with people in a human-oriented way.

1.2.2.2 Proving Theorems by Machine

Mathematics has long been crowned the queen of science. The most extensive basic discipline, it mainly studies number and shape in the real world. As an intellectual discipline, mathematical representation is characterized by precision and deliberateness, and is easy to formalize. Therefore, theorem proof by machine became the first target of scientists in pursuit of AI. A mathematician seeking proof for a known theorem is not only required to deduce a hypothesis, but also to have intuitive skills. A mathematician will adroitly exploit his rich professional knowledge guess which lemma to prove first, decide precisely which known theorems to utilize, divide the main problem into several subproblems, and then solve them one by one. If you can turn these tricky and difficult intelligence behaviors into sophisticated but easy mechanical calculations with a computer, you have mechanized mental labor. Much progress has been made in this area [6].

Generally speaking, the hypothesis of a theorem will be transformed into a set of polynomial equations and its conclusion into a polynomial equation. So theorem proving becomes a purely algebraic exercise in proving one polynomial equation conclusively from a hypothesized set of polynomial equations. In particular, we must mention the outstanding contributions made by two Chinese scholars, Professor Hao Wang and Professor Wenjun Wu (honorary president of the China AI Association), in this regard [6].

Deeper research into theorem proving has provided solutions to problems in specific domains where predicate-based logic language, expert systems, and knowledge engineering are required. Applications have been developed in chemical and mineral analysis, medical diagnosis, information retrieval, and deep research on the equivalence between relational database querying and predicate-logic proof [7].

1.2.2.3 Rivalry between Kasparov and Deep Blue

Game playing is typically considered an intelligent activity. In the year after the Dartmouth Symposium, there was worldwide research into computers that could play chess against humans. Major milestones over the past 60 years in the development of human versus machine chess mirror, to some extent, the history of computer AI development.

In 1956, Arthur Samuel of IBM wrote a self-learning, adaptive checkers program. Just like any excellent checkers player, the computer not only "sees" several moves ahead, but also learns from checkers manuals. In 1958, the IBM704 became the first computer to play against humans, with a speed of 200 moves per second. In 1988, the Deep Thought computer defeated the Danish master player Bent Larsen, with an average speed of 2 million moves per second. In 1997, IBM's Deep Blue, a chess-playing computer, shocked the world with two wins, three draws, and one defeat in matches against the reigning world chess champion Garry Kasparov, using a heuristic search technique. In 2001, the German Deep Fritz chess-playing computer defeated nine of the ten top chess players in the world, with a record-breaking speed of 6 million moves per second.

In October 2015, AlphaGo, an artificial intelligence program developed by Google DeepMind, defeated two dan professional player Fan Hui, the European Go champion, by five games to nil. This was the first time a computer program had defeated a human professional Go player, an unprecedented step in the development of artificial intelligence. The program uses a new approach to computer Go that combines Monte Carlo simulation with two deep neural networks: "value networks," to evaluate board positions; and "policy networks," to select moves, which greatly reduces the effective search space. Since defeating the previous world Go champion Li Sedol in a one to four series in March 2016, AlphaGo has ranked number two in the world. In the face of these successive computer victories, the inevitable question became whether, now that the world's chess and Go champions have been defeated, who can say that the computer is not intelligent?

From the perspective of AI, a match by machine versus human, or a robot competition is a way for scientists to demonstrate AI. In solving problems with similar properties and complexity to chess playing, computers can be considered intelligent. A computer playing against a person takes advantage of reasoning and speed, and focuses more on logical thinking; the human brain may rely more on intuition, imagination, and brand-new tactics and strategies, focusing more on thinking in images. The machine versus human match is essentially "human plus machine versus machine plus human"; the computer is on stage while the human is backstage or vice versa. On one hand, a team of experts in their specific fields prestore huge numbers of steps and strategies backstage, so the computer onstage can decide the appropriate response according to the current situation using complex calculations and analysis. The combined wisdom of experts is accumulated in both software and hardware. On the other hand, top human players can enhance their techniques by interacting with computer players and finding the weaknesses of intelligent computers. In some senses, the rivalry between human and machine is an endless process. Today's world of Internet gaming has opened up a new area for human training and entertainment. Statistically speaking, the result could be an endless tie.

1.2.2.4 Thinking Machine

At the beginning, people called a computer an "electric brain," expecting that it would become a thinking machine. However, the internal structures of existing computers differ significantly from those of the human brain. The computer has experienced tremendous development from electric tube, transistor, and integrated circuit to super large-scale integrated circuit, and there has been enormous progress in both manufacture and performance. The speed of its central processing unit, the storage capacity, integrated circuit density, and bandwidth for communication double, or more, every 18 months, in line with Moore's law, but there is still no breakthrough in the Turing model machine, and computers are still equipped with Von Neumann's architecture. On the other hand, a great deal of intelligent software enables small-sized and miniaturized computers, even embedded computers, to acquire more intelligent behaviors akin to those of humans, such as cognition, recognition, and automatic processing. With rapid advances in pattern recognition technology, including image, speech, and character recognition, the computer has been able to achieve extremely intelligent behaviors, such as self-learning, self-adaptation, self-organizing, and self-repairing.

The rise of the Internet of Things has led to the development of many kinds of perceptive machines, recognizing machines, and behavioral machines, which can enhance the thinking ability of machines with AI. Perception, learning, and understanding of characters, images, voices, speech, behaviors, and friendly interaction with human beings, has raised the intellectual level of these machines. They now include engineering sensors, intelligent instruments, printed character readers, manipulators, intelligent robots, natural language composers, intelligent controllers, and so on. Intelligent robots are expected to be the next emerging strategic industry after cloud computing, the Internet and big data.

1.2.2.5 Artificial Life

Life is the foundation of natural intelligence. It took billions of years for life to appear, and to evolve slowly from single cells to multicellular organisms, from simple life forms to complex animals, eventually giving birth to humankind. However, many questions remain unanswered: what is the essence of life? What is the origin of human life? Is it possible to create an artificial organism from inorganic material? Can we explore these questions by studying artificial materials? These are the questions that inspire the relentless exploring by human beings.

In the middle of twentieth century Alan Turing argued that embryo growth can be studied by computer and Von Neumann tried to describe the logic of biological reproduction using computing. In the 1970s, Chris Langton discovered that life, or intelligence, may come from the edge of chaos. This kind of life, such as computer viruses and cell automation, is quite different from carbon-based life and has since been called artificial life. In 1987, the first symposium on artificial life was held at the Los Alamos National Laboratory. Among the 160 participants were life scientists, physiologists, physicists, and anthropologists. At the symposium, the concept of artificial life was proposed and the direction for imitating natural life was pointed out. It was agreed that artificial life is a man-made system to showcase the behavioral features of natural life. Life as we know it is the classical study of biology, whereas life as it could be is a new subject for AI. Self-reproduction and evolution are two important features of life. Creatures found on the earth are only one form of life. Entities with life features can be made artificially, and research based on computers or other intelligent machines can be conducted in such areas as self-organization and self-replication, growth and mutation, complex systems, evolution and adaptive dynamics, intelligent agents, autonomic systems, robots, and artificial brains.

At present, virtual artificial lives mainly take the form of digital lives created by computer software and virtual computer technology; examples include artificial fish and virtual TV anchormen. Artificial life entities are mainly physical beings and tangible artificial life forms, such as robots and machine cats designed and made using computer technology, automation, and biochemistry.

Can artificial life be realized? Darwin's theory of natural selection has solved the problems of evolution and the origin of species, while the problem of the origin of life is still unsolved. Research on artificial life may hold the solution, but so far, there has been no major breakthrough.

1.2.3 Significant Achievements in AI over the Past 60 Years

The contributions of AI to computer science and to information science as a whole can be best illustrated by the number of AI experts receiving Turing Awards, including Marvin Minsky in 1969, John McCarthy in 1971, Herbert Simon and Allen Newel in 1975, and Edward Albert Feigenbaum and Raj Reddy in 1994. In 2011, American scientist Judea Pearl won the Turing Award for his probabilistic and causal reasoning, elevating the study of artificial intelligence with uncertainty to a new level. They have all made remarkable contributions with an admirable success story in their respective fields.

To judge how AI promotes scientific and technological progress and social development from information technology to wider applications pattern recognition, knowledge engineering, and robots appear particularly prominent.

Pattern recognition used to be a parallel discipline to AI. In recent decades, it has made remarkable progress, gradually becoming absorbed into intelligence science as a core content of AI.

Early research on computer pattern recognition focused on modeling. In the late 1950s, Frank Rosenblatt proposed a simplified mathematical model—perceptron—to imitate the recognition process of the human brain. This recognition system was trained to learn how to classify unclassified patterns according to patterns that had already been classified. By the 1960s the statistical decision approach to pattern recognition had greatly advanced, and a series of works on this approach were published during the 1970s. In the 1980s, John Hopfield discovered the association, storage, and computational power of ANNs, providing a new approach to pattern recognition technology and pointing the way to the study of ANN-based pattern recognition.

As the application of IT becomes more pervasive, pattern recognition is becoming more diversified and varied, with insightful research being conducted on granularities at various levels. The latest trend is pattern recognition of biological features, including speech, character, image, and profile recognition. With a dramatically increasing demand for security information, biological identity-recognition technologies have become "hot property." Identity recognition has become an important tool to ensure social and economic security through image division, feature extraction, classification, clustering, and model matching realized by wavelet transform, fuzzy clustering, genetic algorithms, Bayesian theory, and supporting vector machines. Identity recognition through fingerprints and faces has been extensively used by police and customs officials. Signature recognition is becoming important in ensuring the smooth and secure operation of e-finance and e-business.

At the Fifth International AI Conference held in 1977, Edward Feigenbaum, an advocate of knowledge engineering, spoke on "The Art of AI: Themes and Case Studies of Knowledge Engineering" [8]. The term "knowledge engineering" has since spread to every corner of the world.

In 1968, a research group led by Edward Feigenbaum successfully produced the first expert system for mass spectrometry analysis of the molecular structure of organic chemical compounds, DENDRAL. It picked the correct result out of thousands of possible molecular structures based on a given molecular formula and mass spectrograph of the organic chemical compounds. In 1976, they successfully developed a medical expert system, MYCIN, to help doctors diagnose infectious diseases and give advice on treatment with antibiotics. Since the 1980s, expert systems have been developing rapidly globally and their applications have extended to areas such as tax revenue, customs, agriculture, mining, civil aviation, medicine, and finance.

The natural language understanding represented by machine translation is a shining example of the significant achievements in knowledge engineering over the past 30 years. Natural language understanding uses a computer to process, understand, and generate various natural languages familiar to humans so that the machine can communicate with people who use those natural languages. The research covers lexical, syntactic, grammatical, semantic, and contextual analysis, the structure of phrases, case grammar, language data bank linguistics, computational linguistics, quantitative linguistics, and intermediate language and translation assessment.

With the extensive use of databases and networks, people are immersed in massive data, but are still thirsty for the knowledge and data they actually need. Data authenticity and security have become a major concern. However, people's quest for particular knowledge has made it possible for data mining to rise quickly in the field of knowledge engineering. The term data mining made its first appearance in 1989 at the 11th International Joint Conference on Artificial Intelligence held in Detroit, Michigan. At first, it was called "knowledge discovery in database" (KDD), which the author of this book used as the title of a research project to apply for a 1993 National Nature Science Fund. After 1996, a general consensus was reached on the use of data mining, which means the process of digging up valid, up-to-date, and potentially useful knowledge that is both friendly to people and can be understood by machines. Up until now, objects and tools for mining documents, voices, images, and the web have become popular in the information age, with related works and reference books available everywhere. The symposium on KDD, first initiated by the American Association for Artificial Intelligence in 1989, has been run as an annual international conference since 1995, with competitions organized and awards given. It should be particularly noted that, from the very beginning, data mining has been application-oriented. IBM, GTE, Microsoft, Silicon Graphics, Integral Solution, Thinking Machine, Datamind, Urban Science, AbTech, Unica Technologies successively developed some practical KDD business systems like BehaviorScan, Explorer, and MDT (Management Discovery Tool) for market analysis; Stock Selector and AI (Automated Investor) for financial investment; and Falcon, FAIS, and Clonedetector against fraud.

Research on robots began with the manipulator, a machine capable of imitating human behavior. It was a combination of mechanical structures, sensor technologies, and AI. The first generation of remote-controlled manipulator was made in 1948 at the Argonne National Laboratory and 14 years later came the first industrial robot whose operational process could be easily changed through programming. In the 1960s many scholars from the United States and England who regarded robots as carriers of AI researched how to enable robots to complete environment recognition, problem solving, and planning. In the 1980s, the industrial robot sector took a great leap forward, primarily in the auto industry, with robots used for spot welding, arc welding, spraying, loading, and unloading. During the 1990s, assembly machines and soft assembling technology also made rapid growth.

Robot research has gone through three generations. The first being programmable robots who usually "learn" to work in the following two ways: controlled by preset programs; or by the "demonstration-repeating" method, in which robots learn from repeated demonstrations before they can perform their duties, so when there is a change in the task or the environment, demonstrations are needed again. Programmable robots can work at lathes, furnaces, as welders and on assembly lines. Current robots commercialized for practical applications fall into the demonstration-repeating category.

The second generation is self-adaptive robots with corresponding sensors, such as vision sensors, to obtain simple information from their surroundings. Their actions are analyzed, processed, and controlled by the computer inside their body, and consequently they can adapt to small changes in objects.

The third generation is smart robots, with an intelligence more akin to that of humans. As they are equipped with various types of sensors for vision, sound, and touch, they can perceive multidimensional information on a number of platforms. These robots are highly sensitive in that they can perceive precise information on their surroundings, analyze it in real time, and coordinate their actions. They also have a degree of self-learning ability to deal with changes in the environment and to interact with other robots to accomplish various complicated and difficult tasks. The annual Robot Football World Cup has greatly promoted the research on third-generation robots. Recently, many countries, including China, have launched smart car autonomous driving and a variety of home robots. We can foresee that a higher realm in the Internet of Things may be an Internet of robots. In product manufacturing, domestic services, and other fields, the more complicated and skillful robots will become a bridge connecting human activity with the Internet. But computing based on robotic vision and listening, cognition, and communication with natural language remain a problem to be solved.

1.3 Research Methods for AI

In the 60 years of AI research in terms of basic theories and methodologies, a number of dominant academic schools have appeared.

In 1987, in Boston, an international symposium on fundamental AI issues was jointly sponsored by the Massachusetts Institute of Technology (MIT) AI Lab, the National Science Foundation (NSF), and the American Association for Artificial Intelligence (AAAI). They invited AI experts who had made remarkable contributions to AI development and the participants elaborated on their academic ideas and basic theories and models, discussed their sources, and probed their applicabilities and limitations. Following which, many scholars categorized the main AI theories and methodologies proposed at the symposium as symbolism, connectionism, and behaviorism.

1.3.1 Symbolism

Symbolism holds that cognition is the processing of symbols and that the process of human thinking can be described by symbols. In other words, thinking is computation (or cognition is computation). This ideology once constituted the theoretical base for AI.

The symbolist approach is represented by Herbert Simon and Allen Newell. In 1976, they proposed the Physical Symbol System Hypothesis, holding that a physical symbol system is a necessary and sufficient condition for the presentation of intellectual behaviors. Using mathematical methods to study intelligence and look for the forms, patterns, and formulas of intelligent structure enables intelligence to be systemized and formalized as a mathematical symbol and formula. In this way, any information processing system can be viewed as a materialized physical symbol system that uses rule-based memory to obtain, search for, and control knowledge and operators for solving general problems. Any pattern, as long as it can be distinguished from other patterns, is a symbol. For example, different Chinese characters or English letters are different symbols. The task and function fundamental to a physical symbol system is to identify the same symbols and distinguish different ones. Therefore, it is necessary for a system to be able to locate substantial differences between symbols. Symbols can be concrete, such as the patterns of electrons moving in the physical world or of neurons moving in the human brain. Symbols can also be abstract, such as concepts in human thinking.

A complete symbol system should have the following six basic functions:

1. Inputting symbols
2. Outputting symbols
3. Storing symbols
4. Duplicating symbols
5. Creating symbol structures, that is, forming symbol structures in the symbol system by identifying relations between symbols
6. Conditional migration, that is, carrying on with its action process based on existing symbols

A symbol system capable of completing the whole process is a complete physical symbol system (PSS). Humans possess the above six functions, so do up-to-date computers. Therefore, both human and computer are complete physical symbol systems.

The PSS hypothesis states that any system exhibiting intelligence is certain to have the above-mentioned six functions, and that any system with those six functions is capable of exhibiting intelligence. They provide sufficient and necessary conditions for each other. Along with the hypothesis come three deductions:

Deduction 1: Humans possesses intelligence, so he/she is a PSS.
Deduction 2: Computer is a PSS, so it has intelligence.
Deduction 3: Because both human and computer are PSS, we can use a computer to simulate human intellectual activities.

It should be pointed out that such a physical symbol system is merely an external representation of the system at a certain conceptual granularity. It can be constructed at different conceptual granularities. As a PSS the computer can simulate human intelligence on macroscopic, intermediate, and microscopic levels. A relationship between human and machine is established when an external representation of the six functions is realized at the same conceptual granularity and the computer is able to simulate human intelligence. However, at a smaller conceptual granularity, the intelligential mechanism between human and machine might be totally different. For example, when people are trying to solve the same complex problem, although each individual's thinking process may not be the same, they arrive at the same conclusion. When the computer is used for simulation using a programmatic structure different from a human's thinking process, the same result will be achieved. In this sense, the computer can be said to have realized human intelligence. So, symbolism advocates adopt a functional simulation approach to realize artificial intelligence through analyzing the functions possessed by the human cognitive system and then using the computer to simulate these functions.

The basis of the symbolism approach is symbolic mathematics, which is embodied in predicate calculus and revolution principles, and is realized by means of a program designing tool, namely, a logical programming language.

As the mathematical basis for symbolism is continuously being perfected, any mathematical or logical system can be designed into some sort of production-oriented rule system, which can be used to represent most knowledge. This provides a theoretical basis for the development of expert systems for knowledge in different fields in the information age.

Generally speaking, an expert system is a program system with a large amount of professional knowledge and experience. It makes inferences and judgments, and simulates a decision-making process to solve complex problems that once could only be solved by experts. The expert system has the following technological characteristics [9]:

1. *Concealment of mathematical logic*: Expertise is often nonmathematical. You would not expect specialized experts to express their expertise or reasoning procedure in mathematical symbols and formulas. But the expert system can achieve that by building a general purpose mathematical framework and representing each case with rules. This is the natural way for humans to solve problems by concealing their logic of reasoning procedure. Externally, a rule base is a knowledge base. When the problem is given, the expert system will come up with the correct result by using the rules. The result is a resolution of a principle or a proof in mathematical logic where the user cannot see the proving process.

2. *Nonprocedural of program execution*: The expert system requires only the description of a problem while the procedure of problem solving is realized by the inference engine in the system. The nature of problem solving is to seek the solution to a problem by clause conjunction, matching, instantiation, and back tracking. The actual step taken in problem solving is the execution procedure of the program, which is transparent to both knowledge engineer and programmer.

3. *Openness of expertise expansion*: It is the concealment of symbolic logic and the nonprocedural of program execution in the expert system that enables the easy encapsulation of knowledge and easy updating of databases. So the expert system can expand its knowledge continually and is featured by good openness.

4. *Credibility and friendliness*: Almost all expert systems have explanation mechanisms, whose function it is to explain to the user the system's behavior, what its conclusion is based on, and the reason for its choice. Although it is no more than a recurrence of the concrete procedural of the resolution, this greatly enhances the expert system's credibility and friendliness.

1.3.2 Connectionism

After the 1980s, an ANN revolution took place in AI, enabling connectionism to exert a great impact on symbolism. Connectionism maintains that human cognitive activities are mainly based on the activity of neurons in human brains.

The earliest conception of ANNs can be traced backed to 1943 when two American physiologists, W.S. McCulloch and W. Pitts, proposed a mathematical model for neurons and integrating them into a multilayer model called a neural network [10]. In the 1960s the combination of artificial neural cell models and computers made possible the production of a simple perception machine. This machine has a three-layer structure: an input layer of sensory neural net; a linking layer of central neural net; and an output layer of motor neural net. This perception machine is capable of simple letter, image, and speech sound recognition through learning demonstrations. It is trained by being shown many samples and given a reward or a punishment according to its stimulus-response. But it is incapable of recognizing linearly nonseparable patterns.

In the 1970s, the research on perception machines and neural networks stagnated. John J. Hopfield's proposal of a brand new "Hopfield neural network" in 1982, which successfully solved the salesman problem with a calculating complexity of a NP (Nondeterministic Polynomial) type [5], led to the resurgence of neural network research. In 1983, Geoffrey E. Hinton and Terrence Sejnowski developed a neural network model capable of solving the optimization problem of a nonlinear dynamic system. In 1986, David E. Rumelhart and Geoffrey E. Hinton presented the error back propagation learning algorithm of neural networks in *Nature* [11]. In a conventional feed-forward propagation model, input signals are processed from one layer to the next through multiple hidden layers, ending up propagated to the output layer. Each neuron output in one layer only affects the state of the next layer. A back-propagation model belongs to the error correction learning type, with a learning process made up of two parts: input feed-forward propagation and error feed-back propagation. The error is propagated backward when it appears between the input and the expected output during the feed-forward process. During the back propagation, the connection weight values between each layer of neurons are corrected and gradually adjusted to minimize output error [11].

The representatives of connectionism are McCulloch and Hopfield. They opine that to simulate human intelligence, it is necessary to construct a brain model with the help of bionics. They also maintain that the basic unit of human thinking is neural cells, rather than symbols, and that intelligence is the result of interconnected neural cells competing and coordinating. They hold views different from the physical symbol system hypothesis, believing that the human brain is not the same as the computer. That intelligence is the result of competition and cooperation between those neurons that are connected to each other and that computer simulation of the human brain should focus on structural simulation (i.e., the structure of the human physiological neural network). They also think that function, structure, and behavior are closely related and different structures exhibit different functions and behaviors.

The ANN differs from the symbolist approach in that, in a neuron network, knowledge is represented by the value of weights in interconnection between units. These weights can be continuous. The learning rule of the network depends on the active value equations whose variants are these continuous weights. Thus, the basic units of cognitive and intellectual activities described in the learning rule are the variants of discrete subsymbol values rather than the abstract symbols irrelevant to biology. In this way, the ANN has made a great leap forward toward bionics and is further approaching the neuronal constituent of the human brain on a micro scale. Therefore, the proposal of ANN is regarded as a revolution that transforms the research methodology in cognitive science from symbolism to connectionism.

ANN is widely used in many professional areas, especially in pattern recognition, automatic control, and combinatorial optimization. With the emergence of deep learning, multi-layer computing models have developed quickly in recent years, and thereby greatly improved their computing capacity, operating speed, and training effects. Since then various methods and tools for ANNs that simulate the parallel distributed structure of bio-neural systems have been used for scientific research, business applications, and other areas of life. In addition, some dedicated chips, such as GPU and FPGA, have been designed to achieve the hardware implementation of artificial neural networks.

1.3.3 Behaviorism

Control theory has been used for a long time in AI research. Behaviorists hold that intellectual behavior can be described by a "perception-behavior" model, usually called behaviorism. Its development has gone, roughly, through three phases.

The 1940s to the 1960s were the "classical control theory" period, when frequency analysis—based on transfer functions, frequency properties, and the root-locus method—was used to study linear time invariant systems. Theoretical accomplishments in this period, as represented by H.W. Bode and W.R. Evans, solved the single-input/single-output problem in the production process.

From the 1960s to the 1970s, with the rapid development of the computer, was the period of "modern control theory." A higher-order differential equation was transformed into a first-order differential equation to describe a system's dynamic process, referred to as the state-space method. This method can be used to solve multi-input/multi-output problems, thus expanding a steady-state linear system into a time-varied nonlinear system. Representatives of this period were Lev Semenovich Pontryagin, Carl Michael Bellman, and Rudolph E. Kalman.

From the 1970s, control theory developed into a "generalized system theory." On one hand it studied the structural representation, analytic method, and coordination of a generalized system based on control and information. On the other hand it explored and imitated human perception processes and behaviors, developed information processing procedures and biotic function controls. Representatives were Xuesen Qian, K.S. Fu, and G.N. Saridis.

Early behaviorism focused on human intellectual behaviors and functions in the control process, such as self-optimizing, self-adapting, self-adjusting, self-calming, self-organizing and self-learning, and on the development of so-called "animats." Intelligent control systems and intelligent robots were born in the 1980s, pushing behaviorism in AI research to new heights. In 1988, Rodney Brooks, a representative of behaviorism, created a six-legged walking robot based on a perception-behavior pattern and control system that imitated insect behavior. At present, the inverted pendulum control system and RobotCup are the highlights of AI behaviorist research.

Control theory contains an important branch of robotics research that covers robot sensing, optimal arm movements, and methods of action sequence implementation by a robot. As if in a psychological "stimulus-action" mode, a robot will control corresponding actions once a trigger condition is satisfied. If the trigger condition is sufficient and there is an obvious difference between actions, then the robot will react, even without intelligence. The accuracy of the robot's control depends on the integration density of its circuitry, the speed of the components used for the control algorithm, storage capacity, properties of programmable chips, diagnosis and communication, accuracy of the servo, and so on.

Traditional control theory represents the relation between input and output of the controlled object with a transfer function, which demands an accurate mathematical model and is often difficult to fulfill. Some scholars have researched the controlling mechanism for self-learning and self-organizing and then introduced an artificially intelligent technology into the controlling system. Prof. Jerry M. Mendel employed the learning mechanism on his space flight vehicle and put forward the idea of artificially intelligent control. In 1967, C.T. Leodes and Mendel first introduced the term "intelligent control." From the 1970s, K.S. Fu, Gloriso, and Sardis have suggested that intelligent control is a cross-discipline of AI and control, and set up the system architecture of man–machine interactive hierarchical intelligent control. In the 1980s, the growth of microprocessors and built-in systems provided the conditions for the development of intelligent controllers. Technological advances in knowledge expression, knowledge inference in AI research, and technological progress in the design and construction of expert systems have also provided new means for the study of intelligent control systems. In 1992, the National Science Foundation and the Electric Power Research Institute of the United States initiated a joint research program on Intelligent Control. In 1993, the Institute of Electrical and Electronics Engineers (IEEE) set up a Technical Committee on Intelligent Control. In 1994, the IEEE World Congress on Computational Intelligence was convened in Orlando, it integrated fuzzy logic, neural networks, and evolutionary computation, thus greatly enriching the connotation of intelligent control.

Intelligent control refers to a broad category of control strategies for some devices with apery of intellectual characteristics. It is used on controlled objects whose model parameters are nonlinear, uncertain, and time-varied, even on structures that are hard to describe precisely by mathematical methods. As it is difficult to restrict the outside environment with mathematical parameters, the intelligent control needs to be capable of self-organizing, self-learning, and being self-adaptive (i.e., to accomplish intelligent behaviors), as exemplified by the autonomous machine. There are various control methods, including pattern recognition-based learning control, expert system-based rule control, fuzzy set-based fuzzy control, or ANN-based neural control.

The inverted pendulum is another typical example of intelligent control. It is an inherently unstable system with a varied structure that is impossible to describe mathematically. However, it can learn and be trained through a rule system, and is also capable of resisting outside interference. Therefore, control of the inverted pendulum can exhibit self-organization, self-learning, and self-adaptation with intelligence control.

Symbolism, connectionism, and behaviorism are all simulations of human intelligence. Although there have been great achievements over the last 60 years, there is not yet a unified theoretical system of AI. Different schools have different views over fundamental theoretical AI issues, such as its definition, basics, kernels, architectures, and relation with human intelligence. Regardless of the difference in the three aforementioned research strategies, the AI issue they are researching is only one category of human intelligence. From a wider perspective, such as life psychology, brain science, cognitive science, mathematics, or linguistics, and especially uncertain intelligence, there are still many aspects that have not yet been touched upon, such as human emotion, language, imagination, sudden enlightenment, inspiration, instinct, and so on.

1.4 Interdisciplinary Trends in AI

1.4.1 Brain Science and AI

The human brain is considered the most sophisticated and advanced intelligence system in nature. It is the greatest challenge for natural science to uncover the brain's secrets and mankind has traveled a long way in its exploration. As early as 400 BC, Hippocrates, a doctor in ancient Greece, stated that "the brain is the organ of intelligence." In the seventeenth century, Descartes proposed the concept of "reflexion." In the late nineteenth century, Cajal laid a foundation for neuron theory with a chromosome method named after him. In the twentieth century, Pavlov established the theory of conditioned reflexion of advanced nervous activities. In the 1940s, the invention of the microelectrode opened a new page for neurophysiological research to make great strides in understanding nervous activity.

In the 1960s, neuroscience grew quickly and scientists began to investigate nervous phenomena at cellular and molecular levels. Damage-free brain imaging technology provided an unprecedented tool for scientists to study the activities of a live brain and analyze its working mechanisms. Since the 1990s the emphasis has been on integrating research in brain science. In 1989, the United States took the lead by launching a nationwide brain science program and declared the 1990s to be the decade of the brain. Table 1.1 lists the names of brain scientists and neurobiologists who are Nobel Prize winners.

Brain science aims to understand, protect, and create the brain. So far, AI simulates the human brain by computer and tries to create an artificial brain. Therefore, it is inevitable that brain science and AI are interwoven.

Brain science studies the function of the brain and related diseases in a comprehensive way at molecular, cellular, behavioral, and integrated levels and attempts to understand its structural principles. AI researches how to utilize the computer to simulate the human brain's mental activities—such as consciousness, perception, reasoning, learning, thinking, and planning—to solve complex problems formerly solved only by human experts. Therefore, research on the brain is an inevitable prerequisite for AI. The complexity of the human brain lies in the fact that it is an information processing and decision-making system made up of a trillion (10^{12}) neuron cells and a thousand trillion synapses. It is most likely that human cognitive activities reflected in the brain correspond to physiological, chemical, and electric changes. However, life science is unable to establish a definite relation between mental activities and chemical and electric layers in subcells, in other words, how a concept is stored in biological form and what the biological process is when one concept is connected to others. Life science is also not sure what neuron structures determine the occurrence of which cognitive patterns. Therefore, it is the future task of brain science to study the integrated functions from different angles and to push the understanding of nervous activities at cellular and molecular levels. These research subjects include how the brain performs perception, understands natural language, and generates feelings. An understanding of which will greatly boost the development of AI science.

Although the nature of consciousness remains a mystery after years of work, people are still striving to develop mind-reading machines, memory pills, and smart pills through a 10-year project of behavioral science, which is still being implemented. So our efforts to use machines to simulate human intelligence will be in vain. The impact of brain science development on AI is undoubtedly positive. In terms of the relationship between brain science and AI, we must foster an interdisciplinary awareness that can reveal the nature of brain function, prevent and cure brain disease, and create computers with human intelligence. In short, we should raise awareness to understand, protect, and create brain at the same time.

Table 1.1   Nobel Prize Winners Who Are Brain Scientists or Neurobiologists

Year	Country	Name	Area of Study/Discovery
1904	Russia	I.P. Pavlov (1849–1936)	The theory of conditioned reflexes and signals
1906	Italy	C. Golgi (1843–1926)	The structure of the nervous system
	Spain	S.R. Cajal (1852–1934)	
1932	United Kingdom	C.S. Sherrington (1857–1952)	The functions of neurons
	United Kingdom	E.D. Adrian (1889–1977)	
1936	United Kingdom	H.H. Dale (1875–1968)	Chemical transmission of nerve impulses
	Austria	O. Loewi (1873–1961)	
1944	United States	J. Erlanger (1874–1965)	The highly differentiated functions of single nerve fibers
	United States	H.S. Gasser (1888–1963)	
1949	Switzerland	W.R. Hess (1881–1973)	The functional organization of the interbrain as a coordinator of the activities of the internal organs
1963	Australia	J.C. Eccles (1903–1997)	The ionic mechanisms involved in excitation and inhibition in the peripheral and central portions of the nerve cell membrane
	United Kingdom	A.L. Hodgkin (1914–1998)	
	United Kingdom	A.F. Huxley (1917–)	
1970	United Kingdom	B. Katz (1911–2003)	The humoral transmitters in the nerve terminals and the mechanism for their storage, release, and inactivation in the peripheral and central portions of the nerve cell membrane
	Sweden	U.S.V. Euler (1905–1983)	
	United States	J. Axelrod (1912–2004)	

(Continued)

Table 1.1 (*Continued*)   Nobel Prize Winners Who Are Brain Scientists or Neurobiologists

Year	Country	Name	Area of Study/Discovery
1977	United States	R. Guillemin (1924–)	Hypothalamic hypophysiotropic hormone
	United States	A.V. Schally (1926–)	
1981	United States	R.W. Sperry (1913–1994)	The functional specialization of the cerebral hemispheres
2000	United States	A. Carlsson (1923–)	Signal conduction in the nervous system
	United States	P. Greengard (1925–)	
	United States	E.R. Kandel (1929–)	

1.4.2 Cognitive Science and AI

Cognitive science is derived from cognitive psychology. The term "cognitive science" first appeared in 1975 in *Presentation and Understanding; Studies in Cognitive Science,* cowritten by D.G. Bobrow and A. Collins. In 1977, the journal *Cognitive Science* was inaugurated. In 1979, 23 years after the Dartmouth Symposium on AI, the first cognitive science conference was convened at the University of California, San Diego. D.A. Norman, who chaired the meeting, made a speech entitled "Twelve Issues for Cognitive Science" [12], in which he set out the strategic goals for cognitive science.

Cognitive science studies information processing in the course of human cognition and thinking, including research on perception, memory, learning, language, and other cognitive activities. Perception is the process whereby the brain receives external information in the form of sound, light, touch, and smell through the sensory organs. Visual perception plays a particularly important role. Cognition is based on perception, which is the overall reaction of the brain to the objective world. The presentation of perception constitutes the basis for the study of cognitive processes at all levels. Memory can maintain perceptions, using which current responses can be processed on the basis of previous ones. Only through memory can people accumulate experience. Learning is a basic cognitive activity. The neurobiological basis of learning is the flexibility of synapses that link neural cells, research on which is a very active field in brain science. Some people classify learning into three categories: perceptive, cognitive, and semantic. Learning is mainly presented through language, which is carried by sounds and words and governed by grammatical rules. Language is a symbolic system with a highly complex structure, extremely flexible usage, and extensive application. Thinking activities and cognitive exchanges cannot be performed without language.

Psychologist H. Gardner suggests six disciplines that are closely linked to cognitive science, namely, philosophy, psychology, linguistics, anthropology, artificial intelligence, and neuroscience. For there to be breakthroughs in the search for, presentation of, learning, optimization, prediction, planning, decision-making, and self-adaptation of AI, we must be supported by brain science and cognitive science.

1.4.3 Network Science and AI

The twenty-first century is full of complexity. Edward O. Wilson, cofounder of the U.S. Biological Society once said "The greatest challenge today, not just for cell biology and ecology but for all sciences, is the accurate and complete description of complex systems. Scientists have broken down many kinds of systems and understood most of the elements and influencing forces. The next task is to reassemble them, at least in mathematical models that capture the key properties of the entire ensembles" [13]. A complex network is a model describing a complex system. Understanding the qualitative and quantitative features of network topology, called network science, has become a very challenging topic in the research.

Complex networks, which are abstracted from a wide variety of real networks, often have the topological properties of a large number of nodes and a complex structure. In reality, nodes can be different types of subjects that interact with each other. The research of complex network emphasizes system structure and the analysis of system function based on that structure. The networking and complexity make people no longer pay attention simply to the microdetails of networks, but that they focus more on their overall properties. Natural science has entered a new research era of network science that has penetrated into mathematics, life science, engineering, and many other areas.

We are living in a complex network society (Figure 1.1), there is even a network structure for the psychological world and emotional life of human beings, which plays an important role in the behavior of human beings. Other examples include from the World Wide Web to mobile Internet; from national electricity grid to global traffic network; from metabolism network to neural network; from science collaboration network to social network.

Network topology is an important method to represent knowledge; it can be used to describe a complex system and interaction relationship, which is the basis of the Internet data mining process. It first abstracts the physical network of the real world so as to form a topology structure. However, as the real world is an evolutionary process, random networks, small world and scale-free networks hardly exist in a strictly mathematical meaning, and there is no firm border in topology networks. Therefore, the basic problems for network topology as a means of knowledge representation are as follows: how to generate a network topology to fit real networks; how to generate complex networks; how to simplify and abstract complex networks using typical network models containing uncertain growth, duplication, and variation. During the process of using network topology to represent

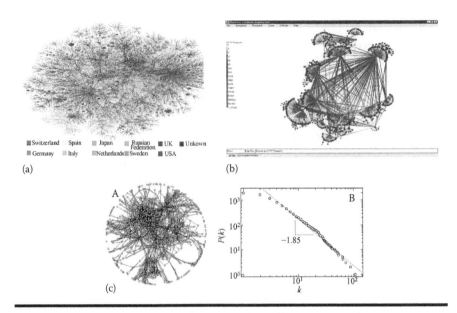

Figure 1.1 We live in a networking society: (a) According to Cheswick, routers can be seen as nodes, fiber connections as edges, different colors as different countries in Internet. (From Cheswick, B. et al., Mapping and visualizing the Internet, In: *Proceedings of Annual Conference on USENIX Annual Technical Conference*, San Diego, California, 2000.) (b) According to Huffaker, the cache can be seen as nodes, the requests as edges, different colors as the volume of HTTP requests. (From Huffaker, J. et al, *Comput. Networks ISDN*, 30(22–23), 2131, 1998.) (c) The ProRally system is mainly used for image processing and simulation of the Playstation2 console. The network topology (A) and degree distribution (B) of more than 1900 classes show that its structure has "small world" and "scale-free." (From S. Valverde, R.V., *DCDIS Series B: Applications & Algorithms (Special Issue on Software Engineering and Complex Networks)*, 14(S6), 1, 2007.)

knowledge, the subject is represented as a node, the relationship between subjects as the edge, and specific physical meaning can be given to the character of edge and node. For example, node character can be used to represent the city size of the traffic network, the throughput of nodes in the Internet, the click-through rate of a website, and the reputation of one person in a social network. Edge can be used to represent the geographical distance between cities in the traffic network, bandwidth between nodes in a communication network, link times between hypertext in the world wide web, and the closeness of social relationships. Furthermore, it is very important to introduce a time character, which can be represented by the wave property in physics, based on which a computing platform can be built to study dynamic behavior in a network, and then to simulate under what critical conditions cascading failure and chain collapse will happen. Overall, using network topology to represent knowledge is an important content of artificial intelligence with uncertainty.

In the study of Internet-based communication behaviors between human beings, the communicating individual can be regarded as a node of subject behaviors, the connection between subjects mapped as edge, and then a typical network topology is achieved. It is very important to study the complex topology of characters in the Internet and in community structures to uncover the rules underlying its superficial disorder, such as the power-law characteristic of node degree distribution, small world features of very small average distances between nodes, the coexistence of vulnerability and robustness, cascading failure and chain collapse, and so on. All these characteristics reveal the general laws of complex networks. Therefore we could say that network science and artificial intelligence are closely combined.

1.4.4 Great Breakthroughs to Be Achieved by Interdisciplinary Research

Science today has developed two significant features: on one hand it is highly divided, there are more and more subdivided disciplines each generating new disciplines and fields; on the other hand, it is highly integrated, in that there is a trend of intercrossing and integration, leading to the emergence of new and cross-disciplinary subjects. Looking back at the history of human development, natural science and social science have developed hand-in-hand, both being a crystallization of the human knowledge system. With the development of human civilization, natural science is more and more closely linked with human activities. The nature of science determines that natural science and social science will be more and more closely and extensively integrated. During the development of modern natural science and social science, each had a strong desire to acquire new scientific approaches from the other. Especially today, with rapid progress in science and technology, the development of natural science is highly reliant on the support of social science.

The universal cross-disciplinary trend is best demonstrated in AI, which is the crystallization of multiple natural and social sciences, such as philosophy, brain science, cognitive science, mathematics, psychology, information science, medical science, biology, linguistics, and anthropology. Breakthroughs made in the research of human intellectual behaviors impact human civilization. However, it is difficult to count on the development of one discipline alone to achieve this. There need to be cooperative efforts among multiple disciplines, and research achievements of one discipline can be shared by another as its research basis or as a means of support. There are numerous examples in this respect. For instance, the application of ANNs in AI is inspired by research results in medical science and biology. Moreover, through arduous efforts, scientists engaging in the human genome initiative have constructed the DNA sequence of human beings and many other creatures. However, without automatic instruments processing large number of samples, and without high-speed computers processing, storing, comparing, and restoring the data, it would have been impossible to accomplish the work of sequencing the 3 billion base pairs of the human genome in so few years. It can be visualized that, as a source of innovative thinking, discipline crossing and integration will foster greater scientific and technologic achievements and inevitably lead to significant breakthroughs in AI.

References

1. K. Popper, *Objective Knowledge: An Evolutionary Approach*, Oxford University Press, New York, 1972.
2. D. Li, C. Liu, Y. Du et al., Artificial intelligence with uncertainty, *Journal of Software*, 15(11), 1583–1594, 2004.
3. H. Wu and L. Cui, *ACM Turing Award (1996–1999). The Epitome of Computer Phylogeny*, High Education Press, Beijing, China, 2000.
4. S. Ma and J. Wang, Review and preview of AI technology. In: *Chinese Academy of Science, High Technology Development Report 2000*, Science Press, Beijing, China, pp. 102–108, 2000.
5. J. J. Hopfield, Neural networks and physical systems with emergent collective computational abilities, *Proceedings of the National Academy of Science of the United States of America*, 79, 2554–2558, 1982.
6. W. Wu, Mathematics mechanization. In: X. Yu and Y. Deng, eds., *The Charming of Science*, Preface, Science Press, Beijing, China, 2002.
7. D. Li and D. Liu, *A Fuzzy PROLOG Database System*, John Wiley & Sons, New York, 1990.
8. A. Barr and E. A. Feigenbaum, The art of AI: Themes and case studies of knowledge engineering. In: *Proceedings of the Fifth International Joint Conference on Artificial Intelligence*, Cambridge, MA, Vol. 5, pp. 1014–1029, 1977.
9. Y. Lin, B. Zhang, and C. Shi, *Expert System and Its Applications*, Tsinghua University Press, Beijing, China, 1988.
10. W. S. McCulloch and W. Pitts, A logical calculus of ideas immanent in nervous activity, *Bulletin of Mathematical Biophysics*, 5, 115–133, 1942.
11. D. E. Rumelhart, G. E. Hinton, and R. J. Williams, Learning internal representations by backpropagating errors, *Nature*, 323(99), 533–536, 1986.
12. D. A. Norman, Twelve issues for cognitive science, *Cognitive Science*, 4(1), 1–32, 1980.
13. E. O. Wilson, *Consilience: The Unity of Knowledge*, Vintage, New York, 1999.
14. B. Cheswick, H. Burch, and S. Branigan, Mapping and visualizing the Internet, In: *Proceedings of Annual Conference on USENIX Annual Technical Conference*, San Diego, California, 2000.
15. B. Huffaker, J. Jung, E. Nemeth et al., Visualization of the growth and topology of the NLANR caching hierarchy, *Computer Networks and ISDN Systems*, 30(22–23), 2131–2139, 1998.
16. S. Valverde, and R. V. Sole, Hierarchical small worlds in software architecture. *Dynamics of Continuous, Discrete and Impulsive Systems Series B: Applications and algorithms (Special Issue on Software Engineering and Complex Networks)*, 14(S6), 1, 2007.

Chapter 2

Cloud Model: A Cognitive Model for Qualitative and Quantitative Transformation

Natural languages, particularly written languages, are not only powerful tools for human thinking, but also the fundamental difference between the intelligence of humans and that of other creatures. Human brains are the carriers of human intelligence and computers the carriers of artificial intelligence. If AI, which is based on quantitative calculation, could use natural language as a way of thinking, then we would have a bridge between the two carriers, performed by qualitative and quantitative formalization.

2.1 Starting Points for the Study of Artificial Intelligence with Uncertainty

2.1.1 Multiple Starting Points

During the 60 years of AI development and growth, people have been seeking research techniques from different perspectives. The "logical school" conducts research at the symbolic level and uses symbols as the elements of cognition, opining that intelligent behaviors are performed by symbolic operations.

The logicians emphasize heuristic search and reasoning. The "biometric school" investigates the neural constructional level and regards neurons as the basic cognitive elements. These researchers invented artificial neural networks and emphasized structural simulation. Intelligence, as they explain it, is the result of competition and cooperation between neurons. Their research subjects include neural properties, structural topology, learning methods, nonlinear dynamics of the network, and adaptive cooperative behaviors. Scholars of the "control school" believe that there cannot be intelligence without feedback. They emphasize the interaction between a system and its environment. The intelligent system acquires information from the environment by sensations and affects the environment by its actions.

As a matter of fact, mathematical axioms and formulas depend on natural language to describe their background and conditions. In the process of studying mathematics, people are used to focusing on formalized symbols and formulas, which are becoming more and more complicated, but often neglect the fact that natural language still plays an important role in the descriptive proof, involving nonproof and uncertainty, which is still fundamental in the axiom system.

From the linguistic perspective, natural language is the essential embodiment of human intelligence, so it has no substitute. No science or technology can exist in the world without being described by natural language. As a result, natural language is an inevitable perspective to research on human intelligence. The foundation of AI with uncertainty lies in the research on the uncertainties of natural languages and their formalization.

2.1.2 *Keeping in Mind Concepts of Natural Languages*

Since natural languages are the carrier of human thought, the development of any natural language becomes a cultural phenomenon. Hence, as the basic unit of natural languages, a linguistic value or concept is essential in human thought.

Generally speaking, philosophers and cognitive scientists consider that the objective world involves physical objects, and the subjective world, in terms of cognitive units and the objects they point to, reflects the inherent and extrinsic relationship between the two worlds. Every thinking activity points to a certain object, and the subjective sensation of that object is perceived due to its existence. Interactions between humans and the environment, including all body motions, sensations, and so on, are reflected in neural activities. Objective features, such as shape, color, sequence, and subjective emotional states, are categorized by the human brain. A new level of representation is then generated based on the categorization result. This may be called an "abstract concept," which is represented by linguistic values. Thereafter, the brain will select a concept and activate related words, or in reverse, extract the corresponding concept out of words received from other people. The linguistic value, directly connected with the concept, can "condense" the cognition and reclassify the objective world. It reduces the complexity of conceptual structures to an acceptable degree. Thus, the concept plays a key role in processing sensations on objects and phenomena.

As human beings keep evolving, natural language is also developing. Human beings, as the inventors of languages, utilize language symbols in the process of thinking. In human evolution, concept-based languages, theories, and models have become tools that help people to describe and understand the objective world.

In natural languages, concept is usually represented by a linguistic value (i.e., the word). In this book, we use linguistic value, word, and concept interchangeably. To this extent, soft computing, or computing with words, or thinking with natural language, all have the same meaning.

2.1.3 Randomness and Fuzziness in Concepts

There are many approaches to the research of uncertainties, among which probability theoretics was the earliest and is the most mature. It originated from research on the inevitability and occasionality of event occurrence, and consists of three branches, namely, probability theory, stochastic study, and mathematical statistics. Then came theories of fuzzy set and rough set, which start from the uncertainty of an event, and membership function and upper/lower approximate sets are introduced. Uncertainty is more often measured by entropy, such as entropy in thermodynamics and information entropy. Chaos and fractal approaches are also used in the research of uncertainty in certain systems. All these approaches are applied to study uncertainty from different perspectives.

It should be said that randomness and fuzziness are two fundamental characteristics of uncertainty, but the research on the correlation between them has not drawn enough attention.

In the fuzzy set, membership functions are applied to describe the extent of uncertainty. Therefore, compared with the assumption of either this or that in probability theory, it is a great progress in understanding. However, the determination of membership functions is often accompanied by subjective thought, as it is either dependent on the experience of experts or acquired by statistical method.

Chinese scholars, such as Nanlun Zhang, calculated the membership degree of the fuzzy concept "youth" in a statistical approach. This large-scale experiment is a classic one for computation of membership functions. Zhang selected 129 valid candidates in Wuhan who independently gave an age range to define the term youth. The data were grouped and the relative frequencies were calculated. Within each group, the median was applied to represent the membership frequency (i.e., the membership value at each point). The results are shown in Table 2.1.

This experiment reflects the differences of subjective cognitions in different environmental conditions in the process of membership acquisition, and thus the acquired membership must involve these differences, which can be explained by probability theoretics.

Table 2.1 Membership Degree of "Youth"

Age	14	15	16	17	18	19
Membership	0.0155	0.2093	0.3953	0.5194	0.9612	0.9690
Age	20	21	22	23	24	25
Membership	1	1	1	1	1	0.9922
Age	26	27	28	29	30	31
Membership	0.7984	0.7829	0.7674	0.6202	0.5969	0.2171
Age	32	33	34	35	36	37
Membership	0.2171	0.2093	0.2093	0.2093	0.0078	0

In probability theoretics, the most fundamental axiomatic presumption is the law of excluded middle (i.e., the sum of probabilities of event A and non-A is strictly equal to one). However, in the concepts used in natural languages this is not necessarily the same. For example, for the same person, the sum of probabilities of events (i.e., the statement that he or she is a youth, and the one that he or she is not a youth) is not necessarily equal to one. In the discussion of possibility, measurement of fuzziness, membership, and possibility measurement are different and it is difficult to clarify the differences between them. In logic theory, the terms "and" and "or" are distinctive, but when used in natural languages as a daily expression, they are difficult to be distinguished. While using uncertain words, such as about, maybe, probably, sometime, usually, around, and so on, people do not care whether they are used to express fuzziness or randomness, especially in cases of emotional and mental descriptions.

Actually, concepts in natural languages are qualitative. To understand the uncertainties in natural language, we do not need to conduct research from the perspectives of randomness and fuzziness. When comparing qualitative to quantitative the gap between randomness and fuzziness is not so restricted and weak. Therefore, we have proposed a more significant approach to establish a model for transformation between qualitative concept and quantitative description.

2.2 Using Cloud Models to Represent Uncertainties of Concepts

It is more natural and more generalized to use concepts to describe the uncertainty between variables than to use mathematics. So how do we express qualitative knowledge with natural languages? How do we realize the inter-transformation between qualitative and quantitative knowledge? How do we represent the soft-reasoning ability in thinking with languages? To answer these questions, we introduce an important cognitive model for qualitative–quantitative transformation that can show the uncertain mechanism during such transformations.

2.2.1 Cloud and Cloud Drops

Let U be a universal set denoted by precise numbers, and C be the qualitative concept related to U. If there is a number $x \in U$, which is a random realization of the qualitative concept C, and the certainty degree of x for C (i.e., $\mu(x) \in [0, 1]$) is a random value with stabilization tendency:

$$\mu : U \to [0, 1] \quad \forall x \in U, \quad x \to \mu(x)$$

then the distribution of x on U is defined as a cloud, and every x is defined as a cloud drop [1–5].

The cloud has the following properties:

1. The universal set U can be either one-dimensional or multidimensional.
2. The random in the definition is realized in terms of probability. The certainty degree is the membership degree in the fuzzy set, and it also has a probability distribution. All these show the correlation between fuzziness and randomness.
3. For any $x \in U$, the mapping from x to [0, 1] is multiple. The certainty degree of x on C is a probability distribution rather than a fixed number.
4. The cloud is composed of lots of cloud drops, which are not necessarily in any order in the universal set. A cloud drop is one realization of the qualitative concept in U. The more cloud drops there are, the better the overall feature of this concept is represented.
5. The more probable the cloud drop appears to be, the higher the certainty degree is, hence the more contribution the cloud drop makes to the concept.

For a better understanding of the cloud, we utilize the joint distribution of (x, μ) to express the concept C. In this book, the joint distribution of (x, μ) is also denoted as $C(x, \mu)$.

Cloud is a model described by linguistic values to represent the uncertain relationship between a specific qualitative concept and its quantitative expression. It is introduced to reflect the uncertainties of concepts in natural languages, which can be explained with the classical probability and fuzzy set theories. The cloud, which is capable of reflecting the correlation between randomness and fuzziness, constructs the mutual mapping between quality and quantity.

The cloud model leads the research on intelligence at the basic linguistic value and focuses on the approach to represent qualitative concepts with quantitative expressions. It is more intuitive and pervasive because, in this way, a concept is transformed into a number of quantitative values or, more vividly, into the points in

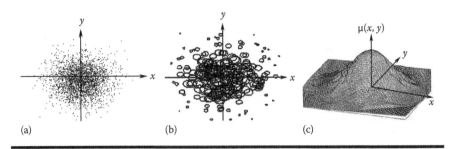

Figure 2.1 Three visualization approaches of the cloud graph: (a) gray-level representation, (b) circle representation, and (c) three-dimensional representation.

the space of the universal set. This discrete transition process is random. The selection of every single point is a random event and can be described by the probability distribution function. Besides, the certainty degree of the cloud drop is fuzzy and can also be described by the probability distribution function. This cloud model is flexible and boundless and its global shape is easier to observe than its local region. These attributes found in the universe are very similar to clouds in the sky in showcasing uncertainties, so we name this mathematical transformation the "cloud."

The cloud can be denoted graphically by the cloud graph. The geometric shape of the cloud is helpful for understanding the transformation between quality and quantity, as in the example of the concept of "the neighborhood at the origin" on a two-dimensional plane. It can demonstrate how to express a concept by cloud drops. There are three visualization methods of the cloud graph [6] in Figure 2.1.

Figure 2.1a shows the position of cloud drops in the two-dimensional space. Every point represents a drop, and its level of gray shows the certainty degree to which this drop can represent the concept. In Figure 2.1b every circle represents a cloud drop whose position and certainty degree can be inferred from the location and size of the circle, respectively. Figure 2.1c illustrates the joint distribution of the drop and certainty degree (i.e., $C(x, y, \mu)$), where (x, y) is the drop's location in two-dimensional space, and μ is the certainty degree, at the third domain, in which a curved surface is formed.

2.2.2 The Cloud's Digital Characteristics

The overall property of a concept can be represented by digital characteristics, which are the overall quantitative property of the qualitative concept. They are of great significance for us to understand both the connotation and the extension of the concept.

In the cloud model, we employ expectation Ex, entropy En, and hyper entropy He to represent the concept as a whole.

Expectation Ex is the certainty measurement of the qualitative concept. It is the mathematical expectation of the cloud drop distributed in the universal set. In other words, it is the point that is most representative of the qualitative concept, or the most classic sample for quantification of the concept.

Entropy *En* is the uncertainty measurement of the qualitative concept. It is determined by both the randomness and the fuzziness of the concept. On the one hand, as a measurement of randomness *En* reflects the dispersing extent of the drops that represent this qualitative concept; on the other hand, it is also a measurement belonging to this qualitative concept, deciding the certainty of the cloud drops that can be accepted by the concept in the domain space. As a result, the correlation between randomness and fuzziness are reflected by using the same digital characteristics.

Hyper entropy *He* is the uncertainty measurement of the entropy (i.e., the second-order entropy of the entropy). As a common sense concept, the higher the degree of general acceptance, the smaller the hyper entropy. As a concept for which it is difficult to form a consensus, the lower the degree of general acceptance, the bigger the hyper entropy. The introduction of hyper entropy provides a means to represent and measure common sense knowledge.

In general, people complete reasoning with the help of natural language, rather than excessive mathematical calculus. The uncertainties in a natural language do not prevent an audience from understanding the content, nor do they keep people from making correct conclusions or decisions. On the contrary, they generate more imaging space for people in their intercommunion and understanding. This is where the beauty of uncertainties shines.

2.2.3 Various Types of Cloud Models

A cloud model is a cognitive model for qualitative-quantitative transformation, which serves as the foundation of cloud transformation, cloud clustering, cloud reasoning, cloud control, and so on. To express a qualitative concept with a quantitative method, we generate cloud drops according to the digital characteristics of the cloud, which is called a "forward cloud generator." The reverse (i.e., from quantitative expression to qualitative concept), is called a "backward cloud generator," which extracts the digital characteristics from the group of cloud drops.

There are various approaches to implement the cloud model that will result in different kinds of clouds based on different distributions, for example, uniform cloud based on uniform distribution, Gaussian cloud based on Gaussian distribution, power law cloud based on power law distribution. There are also a variety of methods for combining cloud models, for example, the symmetric cloud model, the half-cloud model, the combined cloud model, and so on.

The symmetric cloud model represents qualitative concepts with symmetric features, for instance, the qualitative concept "medium height" whose degree distribution is shown in Figure 2.2.

The half-cloud model represents the concept with uncertainty on only one side, for instance, the qualitative concept "fever," reflecting the temperature of a patient whose degree distribution is shown in Figure 2.3.

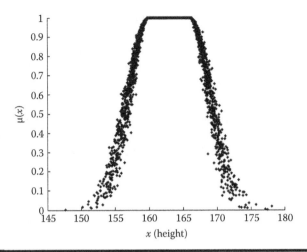

Figure 2.2 Symmetric cloud model.

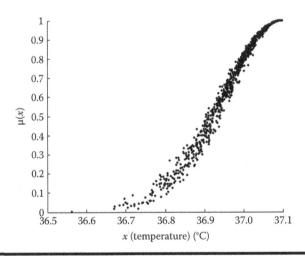

Figure 2.3 Half cloud model.

There are various mathematical characteristics representing uncertainties, and the variation also exists in both natural languages and the real world, for instance, the representation region of some concepts are asymmetric. As a result, we can construct a varying cloud model, for instance, a combined cloud model; Figure 2.4 illustrates a combined cloud model of the qualitative concept "white collar salary."

The cloud model can also be multidimensional. Figure 2.5 shows the joint relationship of the two-dimensional cloud and its certainty degree with the third dimension.

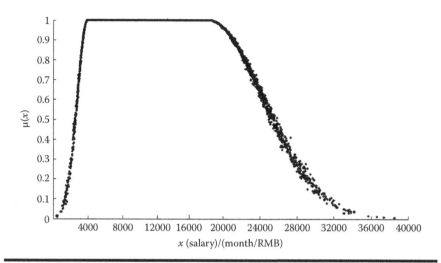

Figure 2.4 Combined cloud model.

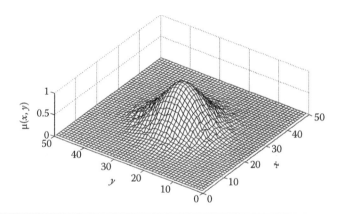

Figure 2.5 Two-dimensional cloud model representing the concept "youth."

2.3 Algorithm of Forward Gaussian Cloud

As one of the most important distributions in probability theory, Gaussian distribution is usually represented by two digital characteristics, the mean and the variance. The bell-shaped membership function is used the most in the fuzzy set, and generally expressed as $\mu(x) = e^{-((x-a)^2/2b^2)}$. The Gaussian cloud is a new model based on both Gaussian distribution and the bell-shaped membership function. The cloud model mentioned herein refers to the Gaussian cloud model if not otherwise specified.

The forward Gaussian cloud is defined as follows.

Let U be a quantitative universal set described by precise numbers, and $C(Ex, En, He)$ be the qualitative concept related to U. If there is a number $x \in U$, which is a random realization of the concept C, and x satisfies $x \sim N(Ex, En'^2)$, where $En' \sim N(En, He^2)$, and the certainty degree of x on C is

$$\mu = e^{-\frac{(x-Ex)^2}{2(En')^2}}$$

then the distribution of x on U is a Gaussian cloud.

2.3.1 Description

2.3.1.1 Forward Gaussian Cloud Algorithm

Input: Digital characteristics (Ex, En, He), and the number of cloud drops n.
Output: n of cloud drops x and their certainty degree μ, that is, drop(x_i, μ_i), $i = 1, 2, \ldots, n$.
Steps:
1. Generate a Gaussian distributed random number En'_i with expectation En and variance He^2, that is, $En'_i = \text{NORM}(En, He^2)$.
2. Generate a Gaussian distributed random number x_i with expectation Ex and variance En'^2_i, that is, $x_i = \text{NORM}\left(Ex, En'^2_i\right)$.
3. Calculate $\mu_i = e^{-\left((x_i-Ex)^2/2En'^2_i\right)}$.
4. x_i with certainty degree of μ_i is a cloud drop in the domain.
5. Repeat steps (1)–(4) until n cloud drops are generated.

This algorithm is applicable to both one-dimensional universal space situations and two or more dimensional situations. The distribution of cloud drops generated by the above algorithm is called Gaussian cloud distribution. A forward Gaussian cloud algorithm can be achieved through integrated circuits and constitute corresponding forward Gaussian cloud generators [7], as shown in Figure 2.6.

Please note that Gaussian cloud distribution is different from Gaussian distribution. The generation of a Gaussian distributed random number is used twice and one random number is the basis of the other, so they are in an iterative relationship. This is the key to the algorithm.

Generally speaking, the variance should not be 0 while Gaussian random numbers are generated. That is why En and He are required to be positive in the algorithm. If $He = 0$, Step 1 will always produce an invariable number En, and as a result

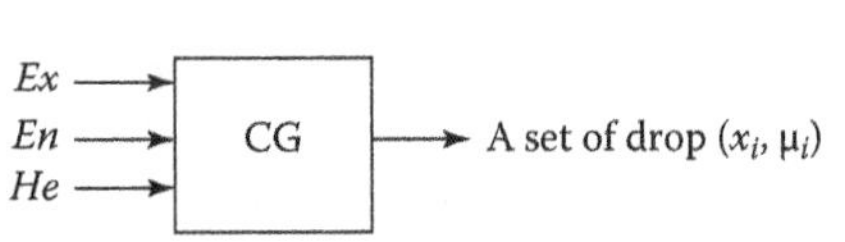

Figure 2.6   Forward cloud generator.

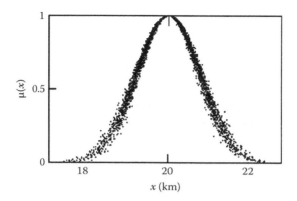

Figure 2.7 The cloud graph $C(x, \mu)$ generated by CG(20, 1, 0.1, 1000).

x will become a Gaussian distribution. If $He = 0$, $En = 0$, the generated x will be a constant Ex and $\mu \equiv 1$. In this sense, certainty is a special case of uncertainty.

For the qualitative concept "about 20 km," with the inputs $Ex = 20$ km, $En = 1$ km, $He = 0.1$ km, and $n = 1000$, the algorithm produces the cloud graph of the joint distribution $C(x, \mu)$ as illustrated in Figure 2.7.

It is not difficult to extend a one-dimensional Gaussian cloud to a two-dimensional case. With the given digital characteristics of a two-dimensional Gaussian cloud [8] (i.e., expectation (Ex, Ey), entropy (Enx, Eny), and hyper entropy (Hex, Hey)), we can generate cloud drops with a two-dimensional forward Gaussian algorithm.

2.3.1.2 Two-Dimensional Forward Gaussian Algorithm

Input: Digital characteristics (Ex, Ey, Enx, Eny, Hex, Hey), generates n cloud drops.
Output: Drop(x_i, y_i, μ_i), $i = 1, 2, \ldots, n$.
Steps:

1. Generate a two-dimensional Gaussian distributed random vector (Enx_i', Eny_i') with expectation (Enx, Eny) and variance (Hex^2, Hey^2).
2. Generate a two-dimensional Gaussian distributed random vector (x_i, y_i) with expectation (Ex, Ey) and variance $\left(Enx_i'^2, Eny_i'^2 \right)$.
3. Calculate $\mu_i = e^{-\left[\frac{(x_i - Ex)^2}{2 En_i'^2} + \frac{(y_i - Ey)^2}{2 Eny_i'^2} \right]}$.
4. Let drop(x_i, y_i, μ_i) be a cloud drop as one quantitative implementation of the linguistic value represented by this cloud. In this drop, (x_i, y_i) is the value in the universal domain corresponding to this qualitative concept, and μ_i is the measured degree of the extent to which (x_i, y_i) belongs to this concept.
5. Repeat steps (1)–(4) until n cloud drops are generated.

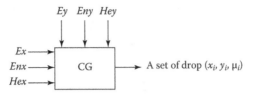

Figure 2.8 Two-dimensional forward cloud generator.

The corresponding two-dimensional forward Gaussian cloud generator is shown in Figure 2.8.

2.3.2 Contributions Made by Cloud Drops to the Concept

In the forward Gaussian cloud model, the cloud drop communities contribute differently to the concept. To understand the forward Gaussian cloud model more deeply, we are going to discuss the different contributions made in different drop communities, based on one-dimensional forward Gaussian cloud model [9].

Within the one-dimensional universal domain, the cloud drop cluster Δx in any small region will make a contribution to the qualitative concept A by ΔA, which satisfies

$$\Delta A \approx \mu_C(x) \times \frac{\Delta x}{\sqrt{2\pi}En}$$

Obviously, the total contribution C made to concept A by all the elements in the universal domain is

$$A = \frac{\displaystyle\int_{-\infty}^{\infty} \mu_C(x)\,dx}{\sqrt{2\pi}En} = \frac{\displaystyle\int_{-\infty}^{\infty} e^{-(x-Ex)^2/2En^2}\,dx}{\sqrt{2\pi}En} = 1$$

because

$$\frac{1}{\sqrt{2\pi}En} \int_{Ex-3En}^{Ex+3En} \mu_C(x)\,dx = 99.74\%$$

the contributive cloud drops for concept A in the universal domain U lie in the domain $[Ex - 3En, En + 3En]$.

Table 2.2 Contributions Made to the Qualitative Concept by Cloud Drops in Different Regions

Name	Interval	Contribution (%)
Key elements	$[Ex − 0.67En, Ex + 0.67En]$	50
Basic elements	$[Ex − En, Ex + En]$	68.26
Peripheral elements	$[Ex − 2En, Ex − En][Ex + En, Ex + 2En]$	27.18
Weak peripheral elements	$[Ex − 3En, Ex − 2En][Ex + 2En, Ex + 3En]$	4.3

As a result, we can neglect the contribution made to concept C by the cloud drops outside the domain $[Ex − 3En, Ex + 3En]$. This is named "the $3En$ rule" of the forward Gaussian cloud.

As listed in Table 2.2, it can be seen by calculation that the cloud drop groups (sample group) located in different sections make different contributions to the concept, so they are referred to as key elements, basic elements, peripheral elements, and weak peripheral elements, as shown in Figure 2.9. Qualitative concepts are often used in natural language, for example, key, basic, peripheral, weak peripheral, and so on, all having predetermined requirements on quantity. As a result we can compute the basic certainty in uncertainty through cloud models. The basic elements,

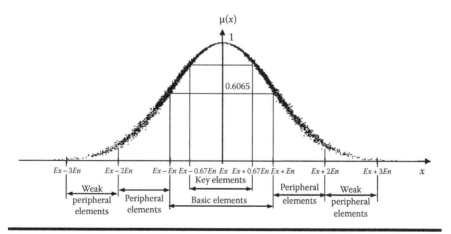

Figure 2.9 Contributions made to the qualitative concept by cloud drops in different regions.

including key elements, cover about 70% of the whole concept, and more than 90% if the peripheral elements are included.

2.3.3 Using Gaussian Cloud Model to Understand the 24 Solar Terms in the Chinese Lunar Calendar

The forward Gaussian cloud model can help people understand a lot of qualitative concepts, which are words or terms in a natural language, for instance, the seasonal concepts of spring, summer, autumn, and winter. The ancient Chinese accumulated abundant experience in farming activities and climate change. They summarized the related knowledge in 24 solar terms, representing their deeper understanding of the four seasonal concepts. We can employ four Gaussian cloud models to represent the four seasonal linguistic values, with 15 as their entropies, and 0.3 as their hyper entropies. The certainty degree of a cloud drop represents the degree to which a specific date belongs to one season.

As recorded in the Spring and Autumn period, there are four dates in the lunar calendar called the vernal equinox, summer solstice, autumnal equinox, and winter solstice. They stand for the expectation of the four qualitative concepts of spring, summer, autumn, and winter, respectively. At these four points, the seasons are spring, summer, autumn, or winter, with a certainty degree of 1. In the cloud graph, the rising trend before each point means that the climate is more like the expected one in that season, and the falling trend after each point means that the season is gradually passing. The symmetry of the graph at these points reflects the transition of the seasons.

At the end of Warring States Period, four other temporal points were added to describe the ending of one season and the beginning of the next, these were "spring begins," "summer begins," "autumn begins," and "winter begins." In the Qin and Han Dynasties, the solar terms became more complete. Two more were added to each of the eight existing solar terms that described the climate in terms of: temperature (e.g., "slight heat," "great heat," "limit of heat," "slight cold," and "severe cold"); moisture (e.g., "white dew," "cold dew," and "hoar frost descends"); rain/snow (e.g., "rain water," "grain rain," "light snow," and "heavy snow"); and the growth of plants (e.g., "excited insects," "pure brightness," "grain fills," and "grain in ear"). Let us take the concept of spring for a detailed discussion. The certainty degrees of rain water and grain rain for the concept of spring are both expected to be 0.1353, while those of excited insects and pure brightness are both 0.6065. The contributions made to spring by the duration of excited insects to pure brightness, from rain water to grain rain, and from spring begins to summer begins are 68.26%, 95.44%, and 99.74%, respectively. The case for summer, autumn, and winter can be made the same way, as shown in Figure 2.10, exhibiting the 24 solar terms represented by the cloud model. It is clear that the points of

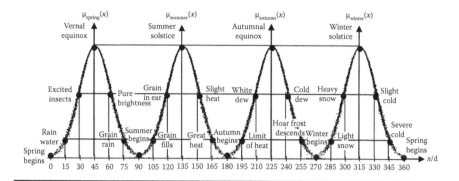

Figure 2.10 Twenty-four solar terms represented by the cloud model.

the 24 solar terms are located on the multiples of *En*, all distributed as the specific points on the expectation curve of the Gaussian cloud, which will be discussed in detail in Section 2.4.3.

2.4 Mathematical Properties of the Gaussian Cloud

Since the Gaussian cloud plays a significant role in the cloud model, it is necessary to analyze its mathematical properties. The forward Gaussian cloud algorithm produces a quantity of cloud drops to constitute a random variable. The degree of certainty of the drops is also a random variable. The two random variables have their own mathematical properties. From a statistical point of view, they both have distribution functions.

2.4.1 Statistical Analysis of the Distribution of Cloud Drops

According to the algorithm of the forward Gaussian cloud, all the cloud drops x will constitute a random variable X, all the En'_i will constitute an intermediate random variable S. The relationship between them is conditional probability.

According to the Gaussian cloud algorithm, S obtains the Gaussian distribution with expectation En and variance He^2, so the probability density function of S is

$$f(s)\frac{1}{\sqrt{2\pi He^2}}\exp\left\{-\frac{(s-En)^2}{2He^2}\right\}, \quad \forall s \in U$$

when $s = \sigma$, X obtains the Gaussian distribution with expectation Ex and variance σ^2. Thus, the conditional probability density function of X is

$$f(x|s = \sigma)\frac{1}{\sqrt{2\pi\sigma^2}}\exp\left\{-\frac{(x - Ex)^2}{2\sigma^2}\right\}, \quad \forall x \in U$$

According to the conditional probability density formula, the probability density function of Gaussian cloud distribution X is

$$f(x) = \int_{-\infty}^{+\infty} f(x|s = \sigma)f(\sigma)d\sigma$$

$$= \int_{-\infty}^{+\infty}\frac{1}{\sqrt{2\pi\sigma^2}}\exp\left\{-\frac{(x - Ex)^2}{2\sigma^2}\right\}\frac{1}{\sqrt{2\pi He^2}}\exp\left\{-\frac{(\sigma - En)^2}{2He^2}\right\}d\sigma$$

Property 2.1

The expectation of Gaussian cloud distribution

$$E(X) = Ex$$

Proof

$$E(X) = \int_{-\infty}^{+\infty} xf(x)dx$$

$$= \int_{-\infty}^{+\infty}\int_{-\infty}^{+\infty} x\frac{1}{\sqrt{2\pi\sigma^2}}\exp\left\{-\frac{(x - Ex)^2}{2\sigma^2}\right\}dx\frac{1}{\sqrt{2\pi He^2}}\exp\left\{-\frac{(\sigma - En)^2}{2He^2}\right\}d\sigma$$

$$= \int_{-\infty}^{+\infty} Ex\frac{1}{\sqrt{2\pi He^2}}\exp\left\{-\frac{(\sigma - En)^2}{2He^2}\right\}d\sigma$$

$$= Ex$$

Property 2.2 [10]

When $0 < He < En/3$, the first-order absolute central moment of Gaussian cloud

$$E\{|X - E(X)|\} = \sqrt{\frac{2}{\pi}}En$$

Proof

$$E\{|X - E(X)|\} = \int_{-\infty}^{+\infty} |x - Ex| f(x) dx$$

$$= 2\int_{Ex}^{+\infty} \frac{x - Ex}{\sqrt{2\pi\sigma^2}} e^{-\frac{(x-Ex)^2}{2\sigma^2}} dx \int_{-\infty}^{+\infty} \frac{1}{\sqrt{2\pi He^2}} e^{-\frac{(\sigma-En)^2}{2He^2}} d\sigma$$

$$= \sqrt{\frac{2}{\pi}} \int_{Ex}^{+\infty} \frac{x - Ex}{|\sigma|} e^{-\frac{(x-Ex)^2}{2\sigma^2}} dx \int_{-\infty}^{+\infty} \frac{1}{\sqrt{2\pi He^2}} e^{-\frac{(\sigma-En)^2}{2He^2}} d\sigma$$

$$= \sqrt{\frac{2}{\pi}} \int_{-\infty}^{+\infty} \frac{|\sigma|}{\sqrt{2\pi He^2}} e^{-\frac{(\sigma-En)^2}{2He^2}} d\sigma$$

$$= \sqrt{\frac{2}{\pi}} En \quad \left(\text{when } 0 < He < \frac{En}{3} \text{ i.e., } \sigma \text{ is positive} \right)$$

Property 2.3

Variance of Gaussian cloud distribution

$$D(X) = E\left\{ \left[X - E(X) \right]^2 \right\} = En^2 + He^2$$

Proof

$$D(X) = \int_{-\infty}^{+\infty} (x - Ex)^2 f(x) dx$$

$$= \int_{-\infty}^{+\infty} (x - Ex)^2 \frac{1}{\sqrt{2\pi\sigma^2}} e^{-\frac{(x-Ex)^2}{2\sigma^2}} dx \int_{-\infty}^{+\infty} \frac{1}{\sqrt{2\pi He^2}} e^{-\frac{(\sigma-En)^2}{2He^2}} d\sigma$$

$$= \int_{-\infty}^{+\infty} \frac{\sigma^2}{\sqrt{2\pi He^2}} e^{-\frac{(\sigma-En)^2}{2He^2}} d\sigma$$

$$= \int_{-\infty}^{+\infty} \frac{(\sigma - En + En)^2}{\sqrt{2\pi He^2}} e^{-\frac{(\sigma-En)^2}{2He^2}} d\sigma$$

$$= \int_{-\infty}^{+\infty} \frac{(\sigma - En)^2}{\sqrt{2\pi He^2}} e^{-\frac{(\sigma-En)^2}{2He^2}} d\sigma + \int_{-\infty}^{+\infty} \frac{En^2}{\sqrt{2\pi He^2}} e^{-\frac{(\sigma-En)^2}{2He^2}} d\sigma + 0$$

$$= En^2 + He^2$$

As shown by the above properties, the cloud drops in a Gaussian cloud model are random variables of expectations Ex, variance $En^2 + He^2$ random variables. This is an important mathematical property of the Gaussian cloud distribution.

Property 2.4 [11]

The third-order central moment of Gaussian cloud

$$E\left\{\left[X - E(X)\right]^3\right\} = 0$$

Proof

$$
\begin{aligned}
E\left\{\left[X - E(X)\right]^3\right\} &= \int_{-\infty}^{+\infty} (x - Ex)^3 f(x) dx \\
&= \int_{-\infty}^{+\infty} (x - Ex)^3 \frac{1}{\sqrt{2\pi\sigma^2}} e^{-\frac{(x-Ex)^2}{2\sigma^2}} dx \int_{-\infty}^{+\infty} \frac{1}{\sqrt{2\pi He^2}} e^{-\frac{(\sigma-En)^2}{2He^2}} d\sigma \\
&= 0
\end{aligned}
$$

Property 2.5 [11]

The fourth-order central moment of Gaussian cloud

$$E\left\{\left[X - E(X)\right]^4\right\} = 9He^4 + 18He^2 En^2 + 3En^4$$

Proof

$$
\begin{aligned}
E\left\{\left[X - E(X)\right]^4\right\} &= \int_{-\infty}^{+\infty} (x - Ex)^4 f(x) dx \\
&= \int_{-\infty}^{+\infty} (x - Ex)^4 \frac{1}{\sqrt{2\pi\sigma^2}} e^{-\frac{(x-Ex)^2}{2\sigma^2}} dx \int_{-\infty}^{+\infty} \frac{1}{\sqrt{2\pi He^2}} e^{-\frac{(\sigma-En)^2}{2He^2}} d\sigma \\
&= \int_{-\infty}^{+\infty} 3\sigma^4 \frac{1}{\sqrt{2\pi He^2}} e^{-\frac{(\sigma-En)^2}{2He^2}} d\sigma \\
&= 3\int_{-\infty}^{+\infty} (\sigma - En + En)^4 \frac{1}{\sqrt{2\pi He^2}} e^{-\frac{(\sigma-En)^2}{2He^2}} d\sigma \\
&= 3\int_{-\infty}^{+\infty} \left[(\sigma - En)^4 + En^4 + 6En^2 (\sigma - En)^2\right] \frac{1}{\sqrt{2\pi He^2}} e^{-\frac{(\sigma-En)^2}{2He^2}} d\sigma \\
&= 9He^4 + 18En^2 He^2 + 3En^4
\end{aligned}
$$

Based on the above five basic properties, *He* reflects how much Gaussian cloud deviates from Gaussian distribution, which can be discussed further as follows.

For a given Gaussian cloud X, a random variable X' can be built very close to a Gaussian variable, where very close means that every central moment of X and X' are equal, and the density function of the Gaussian random variable X' is

$$f(x') = \frac{1}{He\sqrt{2\pi\left(En^2 + He^2\right)}} e^{-\frac{(x'-Ex)^2}{2\left(En^2+He^2\right)}}$$

The expectation, variance, and third-order central moment of X' are Ex, $En^2 + He^2$, and 0, respectively, which are identical to Gaussian cloud distribution; however, the fourth-order central moment of X' is

$$E\left\{\left[X' - Ex\right]^4\right\} = 3\left(En^2 + He^2\right)^2 = 3He^4 + 6He^2En^2 + 3En^4$$

Compared to the fourth-order central moment of Gaussian cloud in Property 2.5, we notice that the fourth-order of central moment of Gaussian cloud distribution X is $6He^4 + 12He^2En^2$ bigger than the fourth-order of central moment of the Gaussian distribution X'.

2.4.2 Statistical Analysis of Certainty Degree of Cloud Drops

Property 2.6

The probability density function of Gaussian cloud drop's certainty degree

$$f(y) = \begin{cases} \dfrac{1}{\sqrt{-\pi\ln y}}, & 0 < y < 1 \\ 0, & \text{otherwise} \end{cases}$$

In other words, the certainty degree distribution of Gaussian cloud drops is independent of the three digital characteristics of Gaussian cloud.

Proof

Based on the forward Gaussian cloud generation algorithm, the certainty degrees of all cloud drops construct random variable Y, and the certainty degree of each drop can be generated by $Y_i = e^{\left(-(X-Ex)^2/2(En_i')^2\right)}$ as a random variable sample.

First, find the distribution function $F_{Y_i}(y)$ of Y_i.

When $y \in (0, 1)$

$$F_{Y_i}(y) = P\{Y_i \le y\}$$

$$= P\left\{ \exp\left[\frac{-(X - Ex)^2}{2En_i'^2} \right]_i \le y \right\}$$

$$= P\left\{ \frac{X - Ex}{En_i'} \le -\sqrt{-2\ln y} \right\} + P\left\{ \frac{X - Ex}{En_i'} \ge \sqrt{-2\ln y} \right\}$$

As

$$X \sim N\left(Ex, En_i'^2 \right)$$

Hence

$$\frac{X - Ex}{En_i'} \sim N(0,1)$$

Then

$$F_{Y_i}(y) = \int_{-\infty}^{-\sqrt{-2\ln y}} \frac{1}{\sqrt{2\pi}} e^{-\frac{t^2}{2}} dt + \int_{\sqrt{-2\ln y}}^{+\infty} \frac{1}{\sqrt{2\pi}} e^{-\frac{t^2}{2}} dt$$

Then the probability density function of Y_i is

$$f_{Y_i}(y) = F_{Y_i}(y)'$$

$$= \frac{1}{\sqrt{2\pi}} e^{-\frac{(-2\ln y)}{2}} \left(\sqrt{-2\ln y} \right)' - \frac{1}{\sqrt{2\pi}} e^{-\frac{(-2\ln y)}{2}} \left(-\sqrt{-2\ln y} \right)'$$

$$= \frac{1}{\sqrt{-\pi \ln y}} \quad y \in (0,1)$$

When $y \ge 1$, $F_{Y_i}(y) = 1$; when $y \le 0$, $F_{Y_i}(y) = 0$. So the probability density of Y_i is

$$f(y) = \begin{cases} \dfrac{1}{\sqrt{-\pi \ln y}}, & 0 < y < 1 \\ 0, & \text{otherwise} \end{cases}$$

Thus, no matter what value En_i' is, the probability density functions of random variable Y_i is the same, though all certainty degrees come from one density function

$$f(y) = \begin{cases} \dfrac{1}{\sqrt{-\pi \ln y}}, & 0 < y < 1 \\ 0, & \text{otherwise} \end{cases}$$

of random variables. $f(y)$ is the probability density function of the random variable Y and the curve is shown in Figure 2.11.

The degree distribution of the cloud drops is independent of the concept's digital characteristics, that is, concept is irrelevant to connotation. Such a fact shows that although there are differences in an individual's cognition of the same concept, the general cognitive law is consistent. This is where the value of the cloud model lies, which is an important discovery.

We can also investigate the probability density function of the joint distribution $C(x_i, \mu_i)$ for x_i within the universal space U.

If U is one-dimensional and $C(x_i, \mu_i)$ is a two-dimensional random variable, we can compute its joint-probability density function. From what we have discussed above, we can see that

$$f_Y(y) = \begin{cases} \dfrac{1}{\sqrt{-\pi \ln y}}, & 0 < y < 1 \\ 0, & \text{otherwise} \end{cases}$$

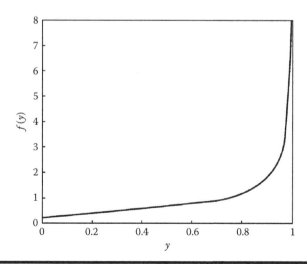

Figure 2.11 The probability density function curve of certainty degree.

to any $\mu = y$,

$$X = Ex \pm \sqrt{-2\ln y}\,En'$$

As $En' \sim N(En, He^2)$, X also obtains Gaussian distribution and its probability density function is

$$f_X\left(x|\mu = y\right) = \begin{cases} \dfrac{1}{\sqrt{2\pi} \times \sqrt{-2\ln y}\,He}\, e^{\dfrac{-\left(x-Ex-\sqrt{-2\ln y}\,En\right)^2}{2\left(\sqrt{-2\ln y}\,He\right)^2}} & Ex \le x < +\infty \\[4mm] \dfrac{1}{\sqrt{2\pi} \times \sqrt{-2\ln y}\,He}\, e^{\dfrac{-\left(x-Ex+\sqrt{-2\ln y}\,En\right)^2}{2\sqrt{-2\ln y}\,He)^2}} & -\infty < x \le Ex \end{cases}$$

Therefore, the joint probability density function of $C(x_i, \mu_i)$ is

$$f_{X,\mu}\left(x, y\right) = f_{\mu}\left(y\right) f_X\left(x|\mu = y\right)$$

$$= \begin{cases} \dfrac{1}{2\pi He \ln y}\, e^{\dfrac{\left(x-Ex-\sqrt{-2\ln y}\,En\right)^2}{4He^2 \ln y}} & 0 < y \le 1,\ Ex \le x < +\infty \\[4mm] \dfrac{1}{2\pi He \ln y}\, e^{\dfrac{\left(x-Ex+\sqrt{-2\ln y}\,En\right)^2}{4He^2 \ln y}} & 0 < y \le 1,\ -\infty < x \le Ex \end{cases}$$

If the universal space becomes more dimensional, the joint-probability density function will be more complicated.

2.4.3 Expectation Curves of Gaussian Cloud

Although the joint distribution $C(x_i, \mu_i)$ has a complex format, the cloud graph generated by the Gaussian cloud algorithm shows obvious geometrical properties. We can study its overall properties by means of both the regression curve and the principle curve.

The regression curve of a Gaussian cloud is formed in this way: for every fixed x_i, the corresponding certainty degree μ_i has expectation of $E\mu_i$, various couples of $(x_i, E\mu_i)$ form the regression curve. The Gaussian cloud has a regression curve as follows:

$$f\left(x\right) = \int_{-\infty}^{+\infty} \frac{1}{\sqrt{2\pi}He}\, e^{-\frac{(y-En)^2}{2He^2}} \times e^{-\frac{(x-Ex)^2}{2y^2}}\, dy$$

Its analytical form is difficult to figure out, but can be approximated by linear estimation.

Every point on the principle curve of a Gaussian cloud is the expectation of every drop of the cloud that is projected to this point. We can also approximate the principle curve of the Gaussian cloud by linear estimation, although the analytical form is not available.

The overall property of the cloud geometric shape is reflected in both the regression curve and the principle curve. The former represents the expectation in the perpendicular direction, and the latter is about the expectation in the orthogonal direction. Can we define an expectation curve in the horizontal direction? We address this question as given in the following paragraph.

As $\mu = e^{-((x-Ex)^2/2En'^2)}$, for an arbitrary $0 < \mu \leq 1$,

$$X = Ex \pm \sqrt{-2\ln\mu}\,En'$$

Since En' is a random variable, X is a random variable located symmetrically on both sides of Ex. We can discuss the case of $X = Ex + \sqrt{-2\ln\mu}\,En'$ and similarly transplant it into the case of $X = Ex - \sqrt{-2\ln\mu}\,En'$.

As $En' \sim N(En, He^2)$, X obtains Gaussian distribution with expectation $EX = Ex + \sqrt{-2\ln\mu}\,En$ and standard deviation $B = \sqrt{DX} = \sqrt{-2\ln\mu}\,He$.

According to $EX = Ex + \sqrt{-2\ln\mu}\,En$, we can get

$$\mu = e^{-\frac{(E(X)-Ex)^2}{2En^2}}$$

the curve

$$\mu(x) = e^{-\frac{(x-Ex)^2}{2En^2}}$$

is so formed: for every fixed μ_j, every point on the expectation curve is the expectation value Ex_i of corresponding drop with the certainty degree μ_j. This curve is called the expectation curve of the Gaussian cloud.

An expectation curve can be utilized both to study the statistical regularity of the spatial distribution of data and to reveal key geometrical properties of the Gaussian cloud. Though the regression and principle curves of a Gaussian cloud may not be expressed analytically, the expectation curve does have an analytical representation. All these curves go through the "center" of drops smoothly, and outline the overall frame of the cloud, showing the framework of the drop set, as all the drops fluctuate randomly around the expectation curve. There are slight differences in the definition of center between the regression, principle, and expectation curves in terms of fluctuation from the perpendicular, orthogonal, and horizontal directions, respectively.

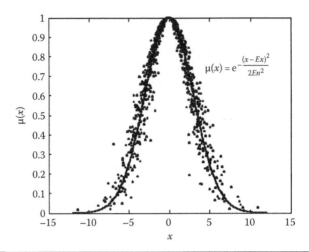

Figure 2.12 The expectation curve of a Gaussian cloud.

The curve in Figure 2.12 is the expectation curve of $C(x, \mu)$ corresponding to the concept $C(Ex = 0, En = 3, He = 0.5)$.

2.4.4 From Cloud to Fog

Entropy *En* and super entropy *He* of a Gaussian cloud are measurements of the uncertainty of a concept. The ratio of entropy and super entropy directly affects the consensus on the represented concept.

1. When *He* = 0, the certainty degree of data samples for the concept is determined and drops of the Gaussian cloud strictly obey the Gaussian distribution, as shown in Figure 2.13a, where the concept extension gathers to form a consensus, representing a mature concept.

2. When 0 < *He* < *En*/3, the certainty degree of data samples for the concept is uncertain and cloud drops of Gaussian cloud gradually deviate from the Gaussian distribution, showing a generalized Gaussian distribution.

    According to the 3*En* rule

$$P\{En - 3He < En' < En + 3He\} = 0.997$$

hence, when 0 < *He* < *En*/3, 99.7% of the cloud drops drip in the region of the outer contour curve

$$y_1 = \exp\left[-\frac{(x - Ex)^2}{2(En - 3He)^2}\right]$$

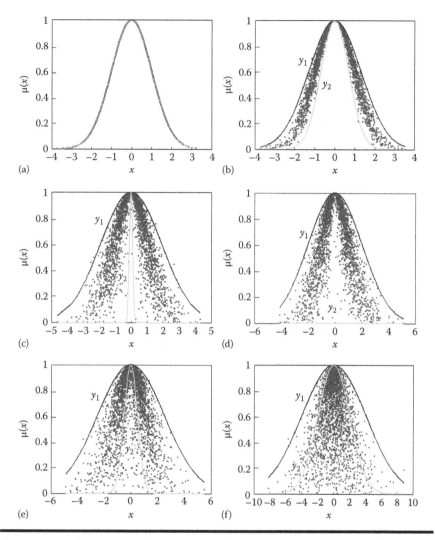

Figure 2.13 The evolution from cloud to fog: (a) *He* = 0, (b) *He* = 0.1*En*, (c) *He* = 0.3*En*, (d) *He* = 1/3*En*, (e) *He* = 0.5*En*, and (f) *He* = *En*.

and inner contour curve

$$y_2 = \exp\left[-\frac{(x - Ex)^2}{2(En - 3He)^2} \right]$$

as shown in Figure 2.13b and c.

3. When *He* = *En*/3, since the exponential of y_2 tends to be negative infinity, the function value tends to be 0, as shown in Figure 2.13d. Cloud drops are more

dispersed in Gaussian cloud. They don't converge into cloud, but show like mist, which means that the concept cannot reach a consensus. The expectation curve of Gaussian cloud is not obvious any more. These drops are called fog.

4. If $He > En/3$, the inner contour curve y_2 starts widening, and part of the cloud drops drip down at the surrounding area outside the curve y_1 and y_2, as shown in Figure 2.13e. As He continues to increase, more and more cloud drops go beyond the region bounded by the two curves in Figure 2.13f. The final trend is that the two function curves tend to coincide $\lim_{He \to \infty} y_2 \to y_1$, then all cloud drops escape from the range of the two curves.

This shows that when a Gaussian cloud represents a concept, $He = En/3$ is a clear border, therefore, a new concept—confusion degree (CD)—is defined as

$$CD = \frac{3He}{En}$$

to measure the degree of dispersion of a concept's extension (i.e., the degree of deviation of Gaussian cloud distribution from the Gaussian distribution). If the CD is 0, then the concept's extension converges to form a consensus, representing a mature concept; if the CD is 1, then the concept's extension diverges, so it is difficult to form a consensus, and the concept is fog. Sometimes we use maturity degree to describe a problem. Confusion degree and maturity are two sides of one problem.

2.5 Algorithm of Backward Gaussian Cloud

A backward Gaussian cloud algorithm can realize the transition from quantitative values to a qualitative concept. It maps a lot of precise data back into the qualitative concept expressed by (Ex, En, He). In essence, the backward Gaussian cloud algorithm is mainly based on a statistical parameter estimation method. There are many parameter estimation methods, such as central moment estimation, maximum likelihood estimation, and least squares estimation. Since the probability density function of the cloud model is an integral expression of nonexplicit analytical formula, it is difficult to use the maximum likelihood estimation method or the least squares estimation method, but the central moment estimation method can be applied. The basic principle for estimating the moment is to use sample moment (which can be calculated by the sample data set) instead of an overall moment, then establish equations on the parameters to be estimated, and finally solve the equations to obtain the sample moment function expression. The easiest and most common method is to use the first-order origin moment to estimate the expectation, and the second-order sample center moment to estimate the variance. The general idea is that if there are K unknown parameters, you can use the K-order sample moment ($K = 1, 2, \ldots$) to estimate the corresponding K-order ($K = 1, 2, \ldots$) population moment, and then use the population moment function involving unknown parameters to calculate the estimated values of these parameters.

This backward cloud algorithm is mainly based on the moment estimation approach. The three algorithms are as follows: a backward cloud algorithm with certainty degree; one without certainty degree based on the first-order absolute central moment and the second-order central moment; and one without certainty degree based on the second-order sample central moment and the fourth-order central moment algorithms.

2.5.1 Description

2.5.1.1 Backward Gaussian Cloud Algorithm with Certainty Degree

Input: Sample points x_i and its corresponding certainty degree y_i, where $i = 1$, 2, ..., n.

Output: Digital characteristics (*Ex*, *En*, *He*) representative of the qualitative concept.

Steps:

1. Calculate the mean of x_i for *Ex*, that is, $\hat{Ex} = \text{MEAN}(x_i)$.
2. For each couple of (x_i, y_i), calculate

$$\hat{En}_i^2 = \frac{-(x_i - \hat{Ex})^2}{2 \ln y_i}$$

3. The random variables En'^2 constructed by all $\hat{En}_i^2$ have the following statistical mathematical property:

$$\begin{cases} E\left(En'^2\right) = \frac{1}{k} \sum_{i=1}^{k} \hat{En}_i^2 = En^2 + He^2 \\ D\left(En'^2\right) = \frac{1}{k-1} \sum_{i=1}^{k} \left(\hat{En}_i^2 - \frac{1}{k} \sum_{i=1}^{k} \hat{En}_i^2 \right)^2 = 2He^4 + 4En^2 He^2 \end{cases}$$

4. Solving equations have

$$\begin{cases} \hat{En}^2 = \frac{1}{2} \sqrt{ 4\left[E\left(\hat{En}'^2 \right) \right]^2 - 2D\left(\hat{En}'^2 \right) } \\ \hat{He}^2 = E\left(\hat{En}'^2 \right) - \hat{En}^2 \end{cases}$$

5. Calculate square root and get $\hat{En}$, $\hat{He}$.

The algorithm is further explained by: $\hat{E}n_i'$ ($i = 1, ..., n$) obtained by the above algorithm in the second step. All $\hat{E}n_i$ ($i = 1, ..., n$) construct the random variable En', all $\hat{E}n_i^2$ construct the random variable En'^2, En' and En'^2 are two intermediate variables. x_i ($i = 1, ..., n$) constructs random variable X, y_i ($i = 1, ..., n$) constructs the random variable Y. Cloud drops variable X and certainty degree variable Y are both output variables. En' obeys Gaussian distribution, its expectation is En, variance is He^2; expectation and variance of En'^2 can be calculated as follows:

1. Because

$$D(En') = E(En'^2) - \left[E(En')\right]^2$$

Hence

$$E(En'^2) = En^2 + He^2$$

2. As

$$D(En'^2) = E(En'^4) - \left[E(En'^2)\right]^2$$

Then

$$E(En'^4) = \int_{-\infty}^{+\infty} En'^4 f_{En'}(En') dx$$

$$= \int_{-\infty}^{+\infty} (En' - En + En)^4 \frac{1}{\sqrt{2\pi He^2}} e^{-\frac{(En'-En)^2}{2He^2}} dEn'$$

$$= \int_{-\infty}^{+\infty} (En' - En)^4 \frac{1}{\sqrt{2\pi He^2}} e^{-\frac{(En'-En)^2}{2He^2}} dEn'$$

$$+ \int_{-\infty}^{+\infty} En^4 \frac{1}{\sqrt{2\pi He^2}} e^{-\frac{(En'-En)^2}{2He^2}} dEn'$$

$$+ 6En^2 \int_{-\infty}^{+\infty} (En' - En)^2 \frac{1}{\sqrt{2\pi He^2}} e^{-\frac{(En'-En)^2}{2He^2}} dEn'$$

$$= 3He^4 + En^4 + 6En^2 He^2$$

Hence

$$D\left(En'^{2}\right) = 2He^{4} + 4En^{2}He^{2}$$

Strictly speaking, backward Gaussian cloud algorithm with certainty degree is the reverse process of the forward cloud algorithm, but because the actual data set will have samples without information on certainty degree, this algorithm is not applicable. Backward cloud algorithm without certainty degree mainly uses mathematical properties of the Gaussian cloud distribution to compute three digital characteristics. According to Properties 2.2 and 2.3 of the Gaussian cloud distribution, the backward cloud algorithm can be achieved based on the first-order absolute central moment and second-order central moment of samples.

2.5.1.2 Backward Cloud Algorithm Based on the First-Order Absolute Central Moment and the Second-Order Central Moment

Input: Sample points x_i, where $i = 1, 2, \ldots, n$.
Output: Digital characteristics of qualitative concept (Ex, En, He).
Steps:

1. Calculate the mean $\hat{Ex} = \left(1/n\right)\sum_{i=1}^{n} x_i$ of x_i, then sample first-order absolute central moment is $\left(1/n\right)\sum_{i=1}^{n}\left|x_i - \hat{Ex}\right|$, sample second-order central moment $S = \left(1/(n-1)\right)\sum_{i=1}^{n}\left(x_i - \hat{Ex}\right)^{2}$

2. According to Property 2.2 of Gaussian cloud distribution, when $0 < He < En/3$

$$\hat{En} = \sqrt{\frac{\pi}{2}} \times \frac{1}{n}\sum_{i=1}^{n}\left|x_i - \hat{Ex}\right|$$

3. According to Property 2.3 of Gaussian cloud distribution: $\hat{He} = \sqrt{S - \hat{En}^{2}}$

The algorithm is simple, $0 < He < En/3$ for most qualitative concepts to be easily satisfied.

According to Properties 2.3 and 2.5 of Gaussian cloud distribution, the third backward cloud algorithm is based on the second-order central moment and the fourth-order central moment of samples.

2.5.1.3 Backward Cloud Algorithm Based on the Second- and Fourth-Order Central Moments of Samples

Input: Sample points x_i, where $i = 1, 2, \ldots, n$.
Output: Digital characteristics of a qualitative concept (Ex, En, He).
Steps:

1. Calculate the mean $\hat{Ex} = (1/n)\sum_{i=1}^{n} x_i$ of x_i, then sample second-order central moment is $c_2 = (1/(n-1))\sum_{i=1}^{n}\left(x_i - \hat{Ex}\right)^2$, sample fourth-order central moment $c_4 = (1/(n-1))\sum_{i=1}^{n}\left(x_i - \hat{Ex}\right)^4$

2. Properties 2.3 and 2.5 of Gaussian cloud distribution yield the following equation groups

$$\begin{cases} c_2 = \hat{He}^2 + \hat{En}^2 \\ c_4 = 9\hat{He}^4 + 3\hat{En}^4 + 18\hat{En}^2\,\hat{He}^2 \end{cases}$$

3. Solve the above equation groups and get:

$$\hat{En} = \sqrt[4]{\frac{9c_2^2 - c_4}{6}}, \quad \hat{He} = \sqrt{c_2 - \sqrt{\frac{9c_2^2 - c_4}{6}}}$$

2.5.2 Parameter Estimation and Error Analysis of Backward Gaussian Cloud

Gaussian cloud distribution is based on, but it is different from, a Gaussian distribution pattern since the parameter estimates are more complex. The estimation accuracy depends on the quality of the sample data set. If the sample data set is far from the Gaussian distribution, there will be a big parameter estimation error, which is natural; if the sample data set is close to a Gaussian distribution the accuracy of the parameter estimation is affected by the size of the sample data set and the deviation from a Gaussian distribution; even if all the sample data sets are made up of a positive Gaussian cloud algorithm, parameter estimation errors will still exist. When the hyper entropy is large the sample's parameter estimation error will be larger.

2.5.2.1 Error Analysis of Ex

All three types of backward cloud algorithms use the mean approach to calculate Ex. According to the Gaussian cloud algorithm, the mean of samples $\overline{X} = (1/n)\sum_{i=1}^{n} x_i$

satisfies Gaussian distribution with Ex and $(1/n)(En^2 + He^2)$ as the expectation and variance. As a result, when we take $\bar{X}$ as the estimation of Ex, the estimation error will be:

$$P\left\{ \left| \bar{X} - Ex \right| < \frac{3}{\sqrt{n}} \sqrt{En^2 + He^2} \right\} = 0.9973$$

Since En is also the parameter to be estimated, we can substitute it with the standard variance of the samples, that is,

$$P\left\{ \left| \bar{X} - Ex \right| < \frac{3}{\sqrt{n}} S \right\} \approx 0.9973$$

Furthermore, if the amount of cloud drops is n, the level of prominence is α, the $(1-\alpha)$-confidence interval of Ex is, $(\bar{X} - t_\alpha (n-1)S/\sqrt{n}, \bar{X} + t_\alpha (n-1)S/\sqrt{n})$, that is,

$$P\left\{ \left| \bar{X} - Ex \right| < t_\alpha (n-1) \frac{S}{\sqrt{n}} \right\} = 1 - \alpha$$

where $\bar{X}, S$ are the mean and standard variances of the samples, respectively, as calculated from the drops. The significance level α takes the values of 0.1, 0.05, 0.01, 0.005, and 0.001. $t_\alpha(n-1)$ is the α-level percentile of the t-distribution with $n-1$ degree of freedom, which can be acquired from the probability distribution table or directly computed.

If $n > 200$, t_α is close to a constant for a given α. At this moment, we can figure out the probability with which Ex lies in a certain interval.

Since

$$DX = \frac{1}{n^2} \sum_{i=1}^{n} En_i'^2 = \frac{1}{n} \left(En^2 + He^2 \right)$$

It shows that the error of Ex will decrease if n increases, and it will increase if En and He increase.

2.5.2.2 Error Analysis of En and He

First, we discuss the case where the sample data set is generated by a forward Gaussian algorithm with small hyper entropy. In the process of cloud drop generation, there exists an iterative relationship which causes error transmission. The error of En is magnified by the error of Ex, then the error of He is magnified by the error of En. Such error transmission is a geometric, rather than an algebraic, series.

For a sample data set close to a Gaussian distribution, but with noisy data, it is more difficult to make an error analysis; if the sample data set is away from the Gaussian distribution—for example, a power law distribution—and if Gaussian cloud distribution is used for the parameter estimation, there will be very big error.

2.5.2.3 Determining the Number of Samples under the Condition of Given Errors and Confidence Level

Normally, there should be a considerable amount of data in the samples because there will be a big error if the amount of data is very small; when the sample data is large enough the error can be neglected, even if the sample size increases. Therefore, backward cloud parameter estimation needs an appropriate sample data set size.

The error analysis of Ex shows that if a significance level α is given, in order to make the estimated error of Ex less than Δ, you can ascertain the minimum number of samples required, that is, estimate the required number of samples n.

Usually, En and He are unknown and n can only be calculated with the help of standard variance of the samples. In a real application, it is good enough to set $\alpha = 0.001$. If $t_{0.001}(n-1)(S/\sqrt{n}) \le \Delta$, the error in Ex estimation will not exceed Δ. On the other hand, $t_{0.001}(n-1) < 4.45$, so $n \ge (20S^2/\Delta^2)$, where S^2 is the standard variance of the samples. The confidence degree of the above result is 99.999%.

If En and He are known, and in order to ascertain the minimum number of cloud drops required, only $n \ge (9(En^2 + He^2)/\Delta^2)$ is sufficient.

In the backward cloud algorithm based on a sample's second-order central moment and fourth-order central moment, there is no restriction between He and En, but $En^2 < 0$ or $He^2 < 0$ often occur in the calculation. That is, there is no solution in the real domain.

If we would like to guarantee $\hat{En}^2 = \sqrt[2]{\left(\left(9c_2^2 - c_4\right)/6\right)} > 0$ and $\hat{He}^2 = c_2 - \sqrt{\left(\left(9c_2^2 - c_4\right)/6\right)} > 0$, then we must guarantee $9c_2^2 > c_4 > 3c_2^2$, that is,

$$9\left(\frac{1}{n-1}\sum_{i=1}^{n}\left(x_i - \hat{Ex}\right)^2\right)^2 > \frac{1}{n-1}\sum_{i=1}^{n}\left(x_i - \hat{Ex}\right)^4 > 3\left(\frac{1}{n-1}\sum_{i=1}^{n}\left(x_i - \hat{Ex}\right)^2\right)^2$$

Obviously, whether the above formula is true or not depends on the error of Ex and the number of samples n, which will be judged according to specific circumstances of the samples. In this regard, Prof. Guoyin Wang presented in the literature [12] that data samples can be grouped to obtain the statistical variance and the mean of each group, making them a new set of samples, then calculate the variance and mean among groups, and use simultaneous equations to obtain entropy and hyper entropy. By solving these grouping functions not only are $En^2 < 0$ or $He^2 < 0$ avoided but also a backward Gaussian cloud algorithm is easily formed.

In short, the cloud model is a cognitive model for qualitative and quantitative transformation based on probability and statistics. It is not necessary that the three digital characteristics of *Ex*, *En*, and *He*, used to present the concept, be overly accurate. For example, the concept of "young people," whether its expectation is 18.1 years old or 18.11 years old, does not differ much in human cognition and thinking, which reflects the robustness of the qualitative concept.

2.6 Further Understanding of the Cloud Model

The cloud model serves as the basis for this study of uncertainty in artificial intelligence so, to better understand it, here are two other cases.

2.6.1 Judgment Shooting

The cloud model is an approach to describe uncertainty problems that, to begin with, people will probably compare with statistical approaches or fuzzy theory, and simply regard it as randomness compensated by fuzziness, fuzziness compensated by randomness, second-order fuzziness, or second-order randomness. This shooting example should clarify some possible misunderstandings.

For example, three scholars—one statistician, one fuzzy scholar, and one cloud model researcher—are invited to judge a shooting competition. Three competitors—A, B, and C—each take 20 shots, with the results as shown in Figure 2.14.

The statistician thinks that "on" and "off" are unambiguous. If it is not on/off, it is off/on and there is no middle status. However, for each shot, it is uncertain whether it is on or off, which is called "randomness." The experimental results are random. Let U be the experimental sample space, that is, $U = \{x\}$, x is the event in the sample space. For every result, we introduce the variable u with a value 0 or 1. For every event, u is variable, so u can be treated as a function defined on the sample space, that is, $u(x)$ is a random variable. The overall performance of the

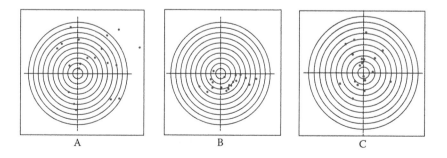

Figure 2.14 The records of shooters A, B, and C.

competitor is measured in terms of the numbers of "on," that is, frequency. For instance, after 20 shots, A got 18 "on" and 2 "off." The probability of "on" for A is 0.9, or a score of 90 if on the 100 scale. All the shots of B and C are "on," so their scores are both 100. As a result, the performances of B and C are equal, and both are better than A.

The fuzzy scholar treats the shooting results differently. He thinks that "on" and "off" are relative rather than absolute, related to the distance from the shot point to the center of the target. There is no precise boundary between "on" and "off." Let x be the element in the sample space $U = \{x\}$. He represents the events as "definitely on" and "definitely off" by 1 and 0, respectively. In this case, some of the elements in the sample space can be expressed in degrees by the numbers between 0 and 1. In this approach, "on" and "off" are represented by the membership degrees of the shot points. The target is divided into 10 concentric circles, which are marked from the inside to the outside with scores from 10 to 9 down to 1. The respective membership degrees are from 1, 0.9 to 0.1. "Off" is marked with 0 and its membership degree is also 0. In this case, even a single shot reveals the skill of the competitor.

Considering the two factors of randomness and fuzziness, the total score can be calculated with the formula score $= \sum_{i=1}^{n} w_i$, where n is the number of shots, w_i is the score of the ith shot. Table 2.3 shows the results for A, B, and C after 20 shots. Their total scores are 80, 136, and 141, respectively. As a result, C's performance is best and A's is worst.

For different shooters, there can be an inherent deviation between the point they aim at and the actual point of impact. For example, from the results of shooter B, his point of impact is expected to be within eight rings, which is very intensive; although the expected mean of shooter C's results is closer to the center of the target than for shooter B, his results are relatively distant from the target. People are more accustomed to using natural language instead of numerical methods to judge shooters' levels. Hitting or missing the target has randomness, and there is an ambiguity in hitting the target, but each point of impact can be treated as a cloud drop. The overall drop characteristics reflect the level of the shooter and we can use a qualitative concept to describe them. The three shooters' results, according to the positions in Table 2.3, have 20 samples in both X and Y axes. Results of the digital characteristics which reflect the levels of the shooters can be calculated by applying the backward Gaussian cloud generating algorithm discussed in this chapter. Table 2.4 illustrates the digital characteristics representing the shooting skills. The expectation (Ex_1, Ex_2) is the coordinates of the average shot point of all the cloud drops, reflecting the aiming skill. The entropy (En_1, En_2) reflects the dispersion degree relative to the average point. The hyper entropy (He_1, He_2) reflects the dispersion degree of the entropy. From these characteristics, we can infer that A shoots a little up and to the right, but his shots are dispersed and not stable; B shoots a little down and to the right, and his shots are concentrated and stable; C shoots closer to the center, but his shots are dispersed and not stable.

Table 2.3 Target Data after 20 Shots for the Three Shooters

	Shooter								
	A			B			C		
Times	X Axis	Y Axis	Rings	X Axis	Y Axis	Rings	X Axis	Y Axis	Rings
1	0.03451	0.09795	9	−0.13753	0.037523	9	−0.42249	−0.19442	6
2	−0.4224	0.48874	4	−0.19442	−0.11633	8	−0.31902	0.51236	4
3	0.84542	−0.4828	1	0.41092	−0.028216	6	−0.17904	0.31782	7
4	0.01840	0.65503	4	0.32156	−0.14648	7	−0.1641	0.58703	4
5	0.05852	0.17723	9	0.72087	−0.10617	3	−0.14554	−0.15835	8
6	0.49159	0.48183	4	−0.15308	−0.17783	8	−0.063297	0.12755	9
7	−0.1455	0.89867	1	−0.15835	−0.28292	7	−0.020185	0.16976	9
8	0.59969	0.73848	1	0.16492	−0.28915	7	0.018401	−0.32156	7
9	0.83737	0.8548	0	0.22966	−0.23945	7	0.034512	−0.13753	9
10	0.34766	0.30481	6	0.088081	−0.0754	9	0.058527	0.72087	3

(Continued)

Table 2.3 (*Continued*) Target Data after 20 Shots for the Three Shooters

Times	Shooter								
	A			B			C		
	X Axis	Y Axis	Rings	X Axis	Y Axis	Rings	X Axis	Y Axis	Rings
11	0.61691	0.20583	4	−0.26154	−0.36671	6	0.20052	0.2542	7
12	−0.0202	−0.7184	3	0.16976	−0.31323	7	0.34766	0.3881	5
13	1.2589	0.5026	0	0.17908	−0.23169	8	0.49159	−0.1531	5
14	0.2005	0.3017	7	0.2542	−0.17036	7	−0.023213	0.24687	8
15	0.79021	0.1643	2	−0.42405	−0.26168	6	−0.0538	0.12586	9
16	0.68509	−0.5015	2	0.022811	−0.27424	8	0.00767	−0.0657	10
17	−0.179	−0.3622	6	0.31782	−0.043327	7	0.0936	−0.2217	8
18	−0.0633	−0.5895	5	0.12755	−0.35071	7	0.18441	−0.46115	6
19	−0.1641	0.1532	8	0.58703	−0.224	4	0.17446	0.00329	9
20	−0.3191	0.5782	4	0.51236	−0.089599	5	0.01318	0.22684	8

Table 2.4   Digital Characteristics of the Shooting Skills, Extracted by the Backward Cloud Generator

Characteristics	Shooter		
	A	*B*	*C*
Ex_1, Ex_2	(0.27, 0.20)	(0.14, −0.19)	(0.016, 0.098)
En_1, En_2	(0.50, 0.48)	(0.29, 0.12)	(0.187, 0.325)
He_1, He_2	(0.19, 0.10)	(0.057, 0.039)	(0.097, 0.077)
Performance	Up-right dispersive unstable	Down-right concentrate stable	Close to the center dispersive unstable

Based on the digital characteristics extracted by the backward cloud generator, we can also use the forward cloud generator to produce more drops for simulating the three sets of shots. Figure 2.15 shows both 20 and 100 shot points of the three shooters, from which we can see that increasing the number of shots reflects their skills more accurately.

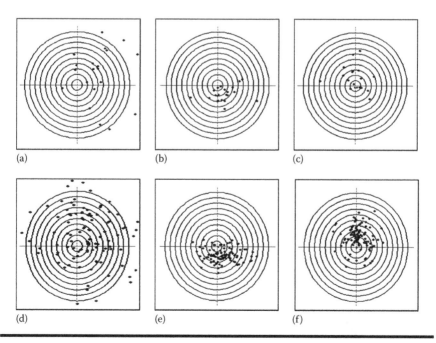

(a)   (b)   (c)

(d)   (e)   (f)

Figure 2.15   Simulation of shot points with forward cloud algorithm: (a) 20 shot points of A, (b) 20 shot points of B, (c) 20 shot points of C, (d) 100 shot points of A, (e) 100 shot points of B, and (f) 100 shot points of C.

2.6.2 Fractal with Uncertainty

As discussed, fractal is an important phenomenon of uncertainty. Fractal geometry is an important tool for the study of chaos phenomena. In nature, among nonlinear complex phenomena, there exist some simple but important self-similar or scale-free features and many natural plant patterns are fractal. The fractal patterns generated by computer simulations are all definite, and do not reflect the uncertainty in the process of natural fractals. Figure 2.16 shows a tree simulation using the classic fractal method. In its growth process, new branches grow to the left or right, forming the left deflection angle α, the right deflection angle β ($0 \leq \alpha, \beta \leq 180°$). The lengths of new branches are divided into the left length ratio l_1 and the right length ratio l_2 ($0 \leq l_1, l_2 \leq 1$). These four parameters can be represented by a tuple of four $\{\alpha, \beta, l_1, l_2\}$. Let a line with a length of L be the initial tree trunk, the fractal tree can be generated according to the following rule R:

1. From the top of the tree, left α, draw the left branch 1: $L \times l_1$, right β, draw the right branch 2: $L \times l_2$.
2. From the top of branch 1, left $\alpha + \alpha$, draw the left branch 3: $L \times l_1 \times l_1$, right $\beta + \beta$, draw the right branch 4: $L \times l_1 \times l_2$. From the top of branch 2, left $\alpha + \alpha$, draw the left branch 5: $L \times l_2 \times l_1$, right $\beta + \beta$, draw the right branch 6: $L \times l_2 \times l_2$.
   $\vdots$

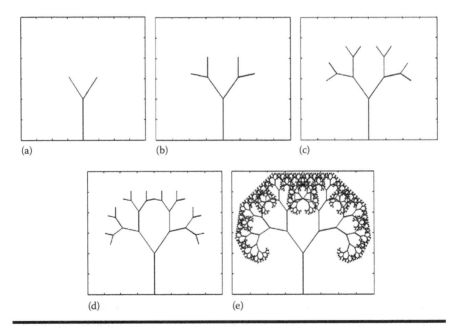

Figure 2.16 Fractal tree: (a) seed, (b) period one, (c) period two, (d) period three, and (e) fractal tree.

⋮

Continue this cycle iteratively until the Mth step. The $2^{M+1} - 1$ short lines will build a binary tree, called a "fractal tree." This is a fractal tree algorithm without uncertainty.

Factors in nature, including sunlight, soil, temperature, or humidity, affect the growth of trees, which as a result, bring about uncertainty during branch growth. Hence, we can use the cloud model as a tool to simulate nondeterministic fractal phenomena in nature. The expectation value of a cloud model can be extended to a parameter set consisting of multiple factors. The function of entropy and hyper entropy to the expectation value can be extended to the function of the whole set of parameters:

$$\alpha_i = CG\left(Ex\_\alpha, En\_\alpha, He\_\alpha\right)$$

$$\beta_i = CG\left(Ex\_\beta, En\_\beta, He\_\beta\right)$$

$$l_{1i} = CG\left(Ex\_l_1, En\_l_1, He\_l_1\right)$$

$$l_{2i} = CG\left(Ex\_l_2, En\_l_2, He\_l_2\right)$$

Let $\{Ex\_\alpha = Ex\_\beta = 40, Ex\_l_1 = Ex\_l_2 = 0.7\}$ be the expectation of the cloud model, $\{En\_\alpha = En\_\beta = 0.05, En\_l_1 = En\_l_2 = 0.1\}$ be the entropy, and $\{He\_\alpha = He\_\beta = 0.001, He\_l_1 = He\_l_2 = 0.01\}$ be the hyper entropy. Using the forward cloud algorithm, a fractal tree with uncertainty can be generated:

$$\alpha_i = CG\left(40, 0.05, 0.001\right)$$

$$\beta_i = CG\left(40, 0.05, 0.001\right)$$

$$l_{1i} = CG\left(0.7, 0.1, 0.01\right)$$

$$l_{2i} = CG\left(0.7, 0.1, 0.01\right)$$

The fractal trees generated according to rules R' based on parameter set $\{\alpha_i, \beta_i, l_{1i}, l_{2i}\}$ are shown in Figure 2.17a through e. The rules R' are as follows:

1. $\alpha_1 = CG(40, 0.05, 0.001)$, $\beta_1 = CG(40, 0.05, 0.001)$, $l_{11} = CG(0.7, 0.1, 0.01)$, $l_{21} = CG(0.7, 0.1, 0.01)$.

   From the top of the tree, left α_1, draw the left branch 1: $L \times l_{11}$, right β_1, draw the right branch 2: $L \times l_{21}$.

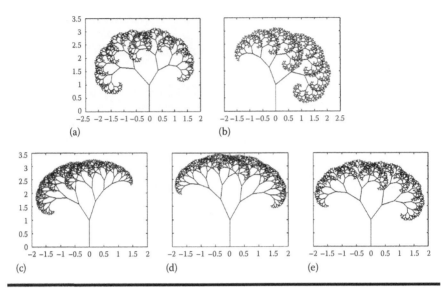

Figure 2.17   Simulating plant variation by virtue of cloud effect on rule set: (a) fractal tree 1, (b) fractal tree 2, (c) fractal tree 3, (d) fractal tree 4, and (e) fractal tree 5.

2. $\alpha_2 = CG(40, 0.05, 0.001)$, $\beta_2 = CG(40, 0.05, 0.001)$, $l_{12} = CG(0.7, 0.1, 0.01)$, $l_{22} = CG(0.7, 0.1, 0.01)$.

From the top of branch 1, left $\alpha_1 + \alpha_2$, draw the left branch 3: $L \times l_{11} \times l_{12}$, right $\beta_1 + \beta_2$, draw the right branch 4: $L \times l_{11} \times l_{22}$. From the top of branch 2, left $\alpha_1 + \alpha_2$, draw the left branch 5: $L \times l_{21} \times l_{12}$, right $\beta_1 + \beta_2$, draw the right branch 6: $L \times l_{21} \times l_{22}$.

Continue this cycle iteratively. Figure 2.17a is the result of 12 steps, a fractal tree with 8191 branches. Obviously, it is different from a fractal tree with certainty.

If the above rules remain unchanged, a repetition of the algorithm will result in different fractal trees, as shown in Figure 2.17b through e.

If the above rules are changed, that is, the droplet previously generated in the iterative process as parameters of the expectations:

1. $\alpha_1 = CG(40, 0.05, 0.001)$, $\beta_1 = CG(40, 0.05, 0.001)$, $l_{11} = CG(0.7, 0.1, 0.01)$, $l_{21} = CG(0.7, 0.1, 0.01)$.

From the top of the tree, left α_1, draw the left branch 1: $L \times l_{11}$, right β_1, draw the right branch 2: $L \times l_{21}$.
2. $\alpha_2 = CG(\alpha_1, 0.05, 0.001)$, $\beta_2 = CG(\beta_1, 0.05, 0.001)$, $l_{12} = CG(l_{11}, 0.1, 0.01)$, $l_{22} = CG(l_{21}, 0.1, 0.01)$.

From the top of branch 1, left $\alpha_1 + \alpha_2$, draw the left branch 3: $L \times l_{11} \times l_{12}$, right $\beta_1 + \beta_2$, draw the right branch 4: $L \times l_{11} \times l_{22}$. From the top of branch 2, left $\alpha_1 + \alpha_2$, draw the left branch 5: $L \times l_{21} \times l_{12}$, right $\beta_1 + \beta_2$, draw the right branch 6: $L \times l_{21} \times l_{22}$.

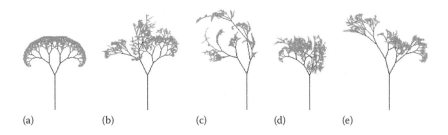

Figure 2.18 Using a cloud model to simulate plant variations: (a) standard fractal tree, (b) fractal tree 1 with variation (c) fractal tree 2, (d) fractal tree 3, and (e) fractal tree 4.

Cycle the following steps:

$$\alpha_i = CG(\alpha_{i-1}, 0.05, 0.001), \quad \beta_i = CG(\beta_{i-1}, 0.05, 0.001),$$

$$l_{1i} = CG(l_{1i-1}, 0.1, 0.01), \quad l_{2i} = CG(l_{2i-1}, 0.1, 0.01)$$

We get more intense fractal trees, as shown in Figure 2.18. Using this method to simulate plant evolution, generation by generation, in the natural environment is a work of great significance.

Then we will get the tree variations.

A new formal method is created by combining a cloud model with fractals, which further shows the randomness contained in the self-similarity of complex objects, thus reflecting the diversity of the uncertainties.

In a different application, the digital characteristics of the cloud model can represent different physical meanings. For example, when the topology of a complex network is generated by a cloud model, the expectation can reflect the average degree of nodes, and the entropy and hyper entropy can reflect the overall uncertainty. To simulate a complex object you can also introduce parameter sets, as in the example of fractal trees.

These two typical cases should facilitate the further understanding of our proposed cognitive model for qualitative and quantitative transformations. In short, the cloud model can be used in a very broad range of applications in the study of uncertainty in humans and nature.

2.7 Universality of the Gaussian Cloud

Randomness and fuzziness used to be the two research topics of probability theory and possibility theory, respectively. Based on these two topics, the cloud model uses probability to explain the membership degree of fuzzy sets. The Gaussian cloud, the most important kind of cloud model, is universal because of the universalities of Gaussian distribution and the bell membership function.

2.7.1 Universality of Gaussian Distribution

Gaussian distribution plays a significant role in academic research and in practical applications of probability theory and the stochastic process. As the most important probability distribution, Gaussian distribution obtains the following form

$$F\left(x;\mu,\sigma^2\right) = \frac{1}{\sqrt{2\pi}\sigma} \int_{-\infty}^{x} e^{-\frac{(x-\mu)^2}{2\sigma^2}} \, \mathrm{d}x$$

Its probability density function is

$$f\left(x;\mu,\sigma^2\right) = \frac{1}{\sqrt{2\pi}\sigma} e^{-\frac{(x-\mu)^2}{2\sigma^2}}$$

where

μ is the expectation representing the most probable value of the random variable

σ^2 is the variance showing the dispersive degree of all possible values

Gaussian distribution exists universally in nature, society, science and technology, and production activities. There are many examples of Gaussian distribution, such as the product quality index under normal production conditions, random errors in measurement, specific features within a biological species, the annual temperature of particular places, and so on.

The central limit theorem theoretically explains the preconditions under which a Gaussian distribution is generated. To simplify the theorem, it can be interpreted as follows. If the value of a random variable is determined by the sum of a quantity of minor and independent random factors, and if every factor has an equally minor influence, that is, there is no dominant factor, then this random variable generally approximates to a Gaussian distribution. For instance, if production conditions (e.g., technique, equipment, manipulation, and raw materials) are normal and stable, the product quality index shall be subject, approximately, to the Gaussian distribution; if the Gaussian distribution is not generated, the production condition must have been changed or is unstable. In practical applications the satisfaction of the Gaussian distribution is judged in this way.

Gaussian distribution is the limited distribution of many important probability distributions. There are a lot of un-Gaussian random variables that are expressed in terms of Gaussian ones. Due to the good properties and the simple mathematical formula of its probability density function and distribution function, Gaussian distribution is widely employed in theoretical research and practical applications.

However, although the emphasis is on the universality of Gaussian distribution, there are still many random phenomena that do not obey Gaussian distribution. Once the determinant factors influence the random phenomena unequally, or are dependent on each other to some extent, the conditions of Gaussian distribution will not be satisfied.

Chairman Mao said in his "On Contradiction": "If there are a number of contradictions in any process, one of them must be the principal and play a leading and decisive role, while the rest occupy a secondary and subordinate position. Therefore, in studying any complex process with two or more contradictions, we must make every effort to find the principal one. Once the principal contradiction is grasped, all problems can be readily solved." Zedong Mao's thoughts, if understood from a scientific point of view, might be a case of power law distribution.

In June 2013, the State Council announced the "Table of General Standard Chinese Characters," the result of a 10-year research project by the Chinese Characters Reform Commission. The table contains a total of 8105 Chinese characters, divided into three levels, with level one for 3500 commonly used words; level two for 3000 less frequently used words; and level three for 1605 words, mainly surnames, place names, and words used in special fields. After comparing the use frequency of many modern Chinese characters, Linguist Youguang Zhou proposed the concept "degressive rate of Chinese characters' efficiency." That is, the coverage for 1000 words with maximum use frequency is 90%. For each additional 1400 words, coverage is increased by only about 10%. This rule applies whether it is a single Chinese character or a word; their random use is not subject to Gaussian distribution because people have a preferential attachment when they use words.

The findings of USA Network scientist Barabási suggest that if the number of nodes is growing in a network topology, and new connections between existing and new nodes come from a preferential attachment, the network so generated is a complex network with distribution subject to a power-distribution, and with a scale-free property.

2.7.2 Universality of Bell-Shaped Membership Function

The concept of membership function is the foundation of fuzzy theory. However, there has not been a widely accepted approach to determining specific membership functions of fuzzy concepts in applications. Instead, they are usually determined case by case and according to experience, as summarized by the following six typical analytical formations,

1. Linear membership function

$$\mu(x) = 1 - kx$$

2. Γ membership function

$$\mu(x) = e^{-kx}$$

3. Convex membership function

$$\mu(x) = 1 - ax^k$$

4. Cauchy membership function

$$\mu(x) = \frac{1}{\left(1 + kx^2\right)}$$

5. Mountain-shaped membership function

$$\mu(x) = \frac{1}{2} - \left(\frac{1}{2}\right) \sin\left\{\left[\frac{\pi}{(b-a)}\right]\left[x - \frac{b-a}{2}\right]\right\}$$

6. Bell membership function

$$\mu(x) = \exp\left[-\frac{(x-a)^2}{2b^2}\right]$$

Although the first three membership functions are continuous, there are some catastrophe points in the membership curves, namely that their first-order derivatives are discontinuous, that is, the left derivative and the right derivative are not equal at these points. If the fuzziness is in nature on a multi scale, the transition in the membership curve should be smooth, so such a sudden change is not allowed. From another aspect, if fuzziness is considered to be existent at both macro and micro scales, concepts should be continuous in terms of all scales, so even higher-order derivatives ought to be continuous. Hence, these three membership functions may only be used in some simple occasions.

Let us discuss the typical example of "youth" as shown in Section 2.1. Nanlun Zhang calculated the membership data of "youth" with a statistical approach. According to the given data, the polynomial curve of the membership function approximates as

$$\mu_0(x) = 1.01302 - 0.00535(x - 24) - 0.00872(x - 24)^2 + 0.0005698(x - 24)^3$$

With the aim of minimizing

$$\left(\frac{\int_a^b \left[\mu(x) - \mu_0(x)\right]^2 \mathrm{d}x}{b-a}\right)^{1/2}$$

instantly, the Cauchy, mountain-shaped, and bell functions are utilized to approximate $\mu_0(x)$. The computation results are, Cauchy membership function is

$$\mu_1(x) = \frac{1}{\left[1 + (x-24)^2/30\right]}$$

Mountain-shaped membership function is

$$\mu_2(x) = \frac{1}{2} - \frac{1}{2}\sin\left\{\left[\frac{\pi}{(37-24)}\right]\left[x - \frac{37-24}{2}\right]\right\}$$

Bell-shaped membership function is

$$\mu_3(x) = \exp\left[\frac{-9(x-24)^2}{338}\right]$$

Calculate the mean squared errors (MSEs) of three membership functions and $\mu_0(x)$ by numerical integration and get 1400 sample points in total; as shown in Table 2.5, we can see the minimum mean square errors is bell.

After comparing the three membership functions with values (X_i, Y_i) to calculate MSEs, the variance of the results are listed in Table 2.6 and the minimum mean square error is still attached to the bell-shaped functions.

It is appropriate to use the bell membership function for such fuzzy concepts as "youth," since their membership functions are extracted statistically, and usually have a shape with a large middle part plus two small side parts. In recent decades, the bell membership function is the most frequently discussed in the fuzzy journals for membership functions. In applications, the membership functions are quite consistent with the bell membership function in many practical fields. As a matter of fact, the sum of low-order terms in the Taylor series of $\mu(x) = \exp[-(x-a)^2/2b^2]$ at the point of a, that is,

$$\mu(x) = 1 - \frac{1}{b^2}(x-a)^2 + \frac{1}{b^4}(x-a)^4 - \cdots$$

Table 2.5 MSEs of Different Membership Function Comparing with $\mu_0(x)$

Membership Function	MSE
Cauchy	0.042181118428255
Mountain-shaped	0.060183795103931
Bell	0.030915588518457

Table 2.6   MSEs of Different Membership Functions Comparing with (X_i, Y_i)

Membership Function	MSE
Polynomial curve fitting	0.083742138813298
Cauchy	0.101253101730926
Mountain-shaped	0.095237099589215
Bell	0.080769137104025

is an approximation of the bell membership function. Compared with other membership functions, the bell membership function enjoys a wider application in many fields.

2.7.3 Universal Relevance of Gaussian Cloud

An understanding of the universality of the Gaussian cloud is quite important. According to the mathematical property of Gaussian cloud as given above, the expectation of random variable X, which is composed of drops is

$$E(X) = Ex$$

Variance is

$$D(X) = En^2 + He^2$$

Nevertheless, the distribution of X is not Gaussian. Thus, how to understand the distribution of X? As we know, Gaussian distribution exists universally in natural phenomena, human society, science and technology, production, and so on. It should have one precondition, that is, if the phenomenon is determined by the sum of a quantity of minor and independent random factors, and if every factor has an equally minor influence, then this phenomenon approximates to a Gaussian distribution. However, there exist many occasions in which there are some relational factors with a great influence. If we simply apply the Gaussian distribution for analysis, we may not reflect what is really going on. For instance, in the shooting case, if every shot is independent, the skill can be represented by the shot points and expressed by a Gaussian distribution. However, every shot may influence the next shot. In this case, a psychological effect needs to be kept in mind, especially in an important competition. Thus, the precondition of Gaussian distribution does not function. To solve this problem, we propose the parameter of hyper entropy to relieve the precondition of Gaussian distribution. For example, we may use *He* to represent the psychological power in different shootings. If *He* is small, the difference between $En^2 + He^2$ and En^2 is small, and thus the shots will be closer to a Gaussian distribution. If *He* is big enough, the shot distribution may be far from Gaussian, revealing instability in the shooting competition.

The condition generating a generalized Gaussian distribution is more relaxed than Gaussian distribution and is more direct and easier than the joint distribution of probability theory in a general sense. When $He = 0$, the Gaussian cloud degenerates Gaussian distribution. In this sense, the generalized Gaussian distribution is more universally significant than a Gaussian distribution.

As we mentioned in Section 2.4, the probability density function of the certainty degree of Gaussian cloud drops is fixed and has nothing to do with the three digital characteristics. According to this mathematical property, we discover an important rule in human cognitive processes, that is, any qualitative concept represented by linguistic values (e.g., "youth," "around 30 degrees") can always be described roughly by Gaussian cloud models. Even though the conceptual or physical meanings, the quantitative distributions in the universal space, and certainty degrees of the cloud drops are all different, statistical distributions of degrees of the cloud drops are generally the same. This means that although the concepts with linguistic values may be differently understood by different persons and in different times, if we neglect the discrete physical meanings, they still reflect the same cognitive regularity in the human brain. There are regularities in the uncertainties of cognition, that is, the consistent common rules within different qualitative representations by different linguistic concepts.

Generally speaking, the two pieces of commonsense knowledge in classical science are repeatability and preciseness. Any unrepeatable temporary phenomenon is not considered scientific. Any phenomenon that cannot be described by a precise mathematical tool is not considered scientific either. However, there is a mathematical approach to overcome such difficulties in the cloud model. It relieves the presumption under which a certain probability distribution is generated and prevents the embarrassment that the membership function in fuzzy theory is artificially determined. This approach has found the regularity of probability density functions for membership functions, which is independent of discrete concepts. The universality of Gaussian distribution and of the bell membership function has laid the theoretical foundation for the universality of the Gaussian cloud model.

No new creative thinking can be the sudden idea of an individual. Rather, it must have deep foundations in practice and also requires long-term accumulation and evolution. In 1923, Bertrand Russell, a famous English logician, started to challenge traditional thought. In his paper entitled "Vagueness" [13], he pointed out that "all traditional logic habitually assumes that precise symbols are being employed. It is therefore not applicable to this terrestrial life, but only to an imagined celestial existence." Russell criticized those who overly emphasized preciseness and pinpointed that "It would be a great mistake to suppose that vague knowledge must be false. On the contrary, a vague belief has a much better chance of being true than a precise one, because there are more possible facts that would verify it." In 1937, Max Black introduced the concept of "consistency profiles," which could be considered as the original status of a membership function. In 1965, L.A. Zadeh, a system scientist, proposed a set of concepts, such as membership degree, membership function, fuzzy set, soft computing, computing with words, and so on. In 1982, a Polish mathematician, Z. Pawlak, introduced the idea of rough set. We have benefited hugely from the application of statistics over 100 years and fuzzy theory for 50 years. The consistency profile on a much higher level is the universality of Gaussian cloud. Today, we could say that it would be a great mistake to suppose that qualitative cognition must be false. On the contrary, qualitative cognition has a much better chance of being true than quantitative, because there are more facts to verify it.

References

1. D. Li, X. Shi, and H. Meng, Membership clouds and membership cloud generators, *Journal of Computer Research and Development*, 32(6), 15–20, 1995.
2. D. Li, X. Shi, and M. M. Gupta, Soft inference mechanism based on cloud models. In: *Proceedings of the First International Workshop on Logic Programming and Soft Computing: Theory and Applications*, Bonn, Germany, pp. 38–62, 1996.
3. D. Li, J. Han, and X. Shi, Knowledge representation and discovery based on linguistic models. In: H. J. Lu and H. Motoda, eds., *KDD: Techniques and Applications*, World Scientific Press, Singapore, pp. 3–20, 1997.
4. D. Li, J. Han, X. Shi et al., Knowledge representation and discovery based on linguistic atoms, *Knowledge-Based Systems*, 10(7), 431–440, 1998.
5. D. Li, Knowledge representation in KDD based on linguistic atoms, *Journal of Computer Science and Technology*, 12(6), 481–496, 1997.
6. D. Li, Y. Wang, and H. Lü, Study on knowledge discovery mechanism. In: Chinese Association for Artificial Intelligence, eds., *The Progress of Artificial Intelligence in China 2001*, Beijing University of Posts and Telecommunications Press, Beijing, China, pp. 314–325, 2001.
7. D. Li, Membership Cloud Generation Method, Membership Cloud Generator and Membership Cloud Control Device, Chinese Intellectual Property Number ZL95 1 03696.3.
8. Z. Yang and D. Li, Two-dimensional cloud model and its application in prediction, *Chinese Journal of Computers*, 21(11), 962–969, 1998.
9. D. Li, Uncertainty in knowledge representation, *The Chinese Engineering Science*, 2(10), 73–79, 2000.
10. C. Liu, A new backward cloud algorithm based on cloud X, *Journal of System Simulation*, 16(11), 2417–2420, 2004.
11. L. Wang, Definition and mathematical properties of Gaussian cloud, Personal Letter, 2011.
12. G. Wang, Backward Gaussian cloud algorithm based on sample grouping, Personal Letter, 2012.
13. B. Russell, Vagueness, *Psychology and Philosophy*, 1(6), 84–92, 1923.

Chapter 3

Gaussian Cloud Transformation

The basic unit of human thinking in natural languages is a concept. The cloud model is a cognitive model for qualitative–quantitative transformation. Concepts are formed by different environments, at different levels, and in different aspects of a problem. Human intelligence activities can be carried out and quickly altered at different granularities, and vice versa.

The implied premise of a backward cloud algorithm, as presented in Chapter 2, is that the entire given data sample corresponds to the same concept, but that algorithm cannot solve multigranularity and multiconcept generation problems across the problem domain. This limits the cognitive abilities of the backward cloud algorithm to generate more concepts and more knowledge from the data sample.

Gaussian transformation (GT), also known as the Gaussian mixture model (GMM), is an important probability and statistics method, and can turn the probability density distribution of the entire problem domain into the superposition of a multiple Gaussian distribution, which provides a multigranularity concept solution guideline.

This chapter, based on cloud model and GT, proposes and implements a Gaussian cloud transformation (GCT), which will convert the data distribution in the problem domain into a number of concepts with different granularities. GCT can be used to address the challenges of generating, selecting, and optimizing a concept's number, granularity, and level in variable granular computing, and to provide a new method for data clustering of big data.

3.1 Terminology in Granular Computing

3.1.1 Scale, Level, and Granularity

Scale, level, and granularity are fundamental terminologies in granular computing that serve as the basis for the research of concept.

Scale refers to the space or time unit of an object or phenomenon, a range of phenomena or processes in space and time, and the window used by people when they observe things, objects, patterns, or processes. In brief, scale is the description of the relative size of the object in the container. In different disciplines, the expression or meaning of scale is also different. In geomatics, cartography, and geography, scale is the ratio of the actual distance to its expression on the map. With the progress of life science and information technology, we now understand humans and nature at different scales, as shown in Figure 3.1. In Figure 3.1, objects being studied in particle physics involve a 10^{-15} m and 10^{-22} s physical scale of magnitude, while astrophysics has brought us to 10^{16} years magnitude, the, so-called, age of the universe. The two arrows facing each other tell us that understanding humans and nature at different scales may present an overall self-similarity. The relationship between intelligence and nature is much closer than ever before, so we need to understand intelligence at different scales and make sure that self-similarity actually exists.

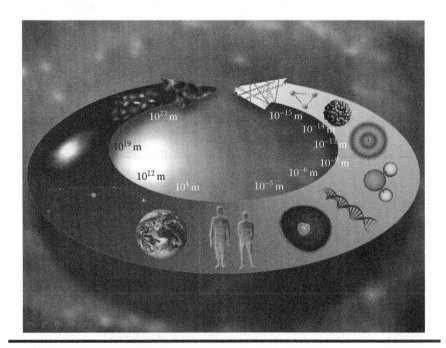

Figure 3.1 Multi-scale views of the world.

The brain is often called a small universe, and cognition is the subjective reflection of the objective world. When dealing with the relationship between different concepts, people used to divide levels into macro, meso, and microscopic to understand the hierarchical relationships and form a concept tree. These macro, meso, and micro levels represent a concept's granularity; the larger the granularity, the wider the scope of the data and the more abstract and macro the concept; the smaller the granularity, the narrower the scope of the data, and the more specific and detailed the concept.

Granularity, originally a physics concept, is a measurement of average particle size, which is used herein to measure the amount of information contained in data. From different concept hierarchies to analyzing and processing data in the domain space, especially for big data, it helps to understand the amount of information in different granularities. If we take the cloud model as the basic model of concept representation, then expectation (*Ex*) is the model nucleus, entropy (*En*) reflects the degree of data dispersion, indicating the granularity, and hyper entropy (*He*) is a measurement of the maturity of the concept.

As Figure 3.2 shows [1], complex systems often have common characteristics, involving many components dynamically interacting and giving rise to a number of levels, or scales, which exhibit common behaviors. As acknowledged, human cognition features the ability to observe and analyze the same phenomenon or problem from different granularities. From the concept of a fine granularity to one with a coarser granularity is the abstraction of information or knowledge, which can simplify the problem, and usually be called data reduction. Observation and the

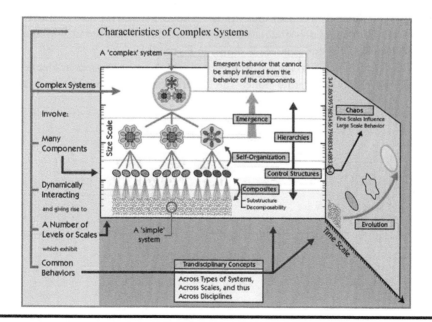

Figure 3.2   Characteristics of complex systems.

analysis of information with coarse granular concepts will find commonalities by ignoring the nuances of a finer granularity. Commonalities are often more profound than personalities and can control at a macroscopic level. Conversely, the observation and analysis of information with a fine granular information can locate more complex personalities, accurately distinguishing differences and differentiating between minorities. Personality is more typical and richer than commonality, but it cannot be fully integrated into the commonality. Through concept promotion, we can find more general knowledge.

3.1.2 Concept Tree and Pan-Concept Tree

The hierarchy of a traditional concept tree, usually expressed by a partially ordered set $(H, \prec)$, where H is a finite set of objects, $\prec$ is a partial order on the H, having non-symmetric and transitive properties. According to the partial order $\prec$, H can be divided into a multiple disjoint collection of objects $H = \cup H_i$, thus we use a tree structure to represent the object hierarchy. For example, according to the age division proposed by the United Nations World Health Organization in 2000, people aged below 44 years are young, people aged from 45 to 59 years are middle-aged, people aged from 60 to 74 years are the young old, and those from 75 to 89 years are the elderly, people aged over 90 years are long-lived. Figure 3.3 is drawn according to such a division method.

Another example is ages of school students, which are usually divided as 3–5 year olds in kindergarten; 6–11 year olds in elementary school; 12–14 year olds in junior high school; 15–17 year olds in high school; 18–21 year olds in university; 22–24 year olds are master graduate students; and 25 year olds are beginning their doctoral studies. The concept tree is shown in Figure 3.4.

Except for the top root node, the other nodes of the concept trees in Figures 3.3 and 3.4 have only one father node. The hierarchical relationship between nodes

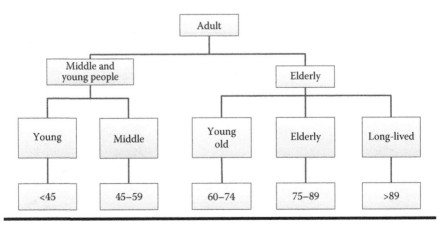

Figure 3.3   Age concept tree by World Health Organization.

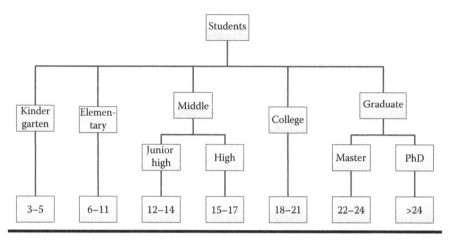

Figure 3.4 Concept tree divided by school age.

is very clear. However, such a concept tree has obvious limitations. For example, the boundaries of numerical intervals corresponding to different concepts are clear, which does not allow for overlaps, so there is a lack of inherent fuzziness in the concept. It is too arbitrary to put 44 and 45 year olds into two different age concepts. This kind of single tree-shaped affiliation relation cannot reflect the uncertainty that a property value may belong to other upper concepts at the same time.

Typically, the formation of a concept tree is related to specific themes, times, and geographic areas, so it is relative. Maybe for scientists 48 years old is still very young. In 1995, the World Health Organization decided that middle-age was from 45 to 65 years old then in 2000 changed it to 45 to 59 years old; in other examples the United States decided middle-age was from 40 to 65 years old, Portugal that it was from 29 to 51 years old, Japan and China from 40 to 60 years old.

In a cognitive process people often cannot form a structured tree with a clear border demarcation, so overlap may exist between concepts, and a low-level concept can belong to more than one high-level concept, this can be called a pan-concept tree and is caused by cognitive uncertainty.

3.2 Gaussian Transformation

Space transformation, or domain transformation, is a commonly used method in scientific research. A complicated problem in one space may become simpler and easier to understand after it is transformed to another space. For example, in physics the Fourier transform can transform a time domain function into a superposition of multiple sine functions in the frequency domain. This transformation is unique, and fast Fourier transform is an important tool in modern engineering applications.

Inspired by the above ideas, and taking into account the universality of Gaussian distribution and the universality of the Gaussian cloud in conceptual representation, is it possible to characterize any probability density distribution of the problem domain by the superposition of a number of Gaussian clouds? To do so it is necessary to introduce the Gaussian transform and its parameter estimation.

Gaussian transform (GT) is the process of transforming probability density distribution into the superposition of a number of Gaussian distributions. Random variable x in the problem domain can be one-dimensional or multi-dimensional, and its mathematical representation is

$$p(x) \rightarrow \sum_{i=1}^{M} \left(a_i G\left(x; \mu_i, \Sigma_i \right) \right)$$

In formula:

$$G\left(x; \mu_i, \Sigma_i \right) = \frac{1}{\sqrt{(2\pi)^d \left| \Sigma_i \right|}} e^{-(1/2)(x-\mu_i)^T \Sigma_i^{-1}(x-\mu_i)}$$

a_i, μ_i, Σ_i, respectively, are the amplitude of ith Gaussian distribution, expectation, and covariance matrix, and satisfies $\sum_{i=1}^{M} a_i = 1$; d is the dimension of data; M is the number of Gaussian distributions.

Usually the expectation maximization algorithm [2] is used to estimate the Gaussian transform parameters. Obviously, error exists in this process and results are not unique.

Furthermore, when solving actual problems, it is frequently unable to get an accurate probability density function, so it may be replaced by a frequency histogram whose statistical properties determine its consistency.

3.2.1 Parameter Estimation of Gaussian Transform

For any frequency distribution of any one-dimensional data, within a certain error range, the mathematical representation of M superimposed Gaussian distributions produced by GT is

$$p(x|\Theta) = \sum_{i=1}^{M} a_i p_i\left(x|\theta_i \right)$$

where

$p_i(x|\theta_i)$ is the Gaussian density function

θ_i is the corresponding parameter, such as expectation or variance

$\Theta = (a_1, \ldots, a_M, \theta_1, \ldots, \theta_M)$ are all parameters to be estimated by GT

The parameter estimation of GT is to evaluate parameter Θ based on the sample x. If $X = (x_1, x_2, \ldots, x_N)$ be the occurred events, and x_i $(i = 1, \ldots, N)$ are mutually independent events, then

$$p(X|\Theta) = \prod_{i=1}^{N} \sum_{j=1}^{M} a_j p_j (x_i|\theta_j)$$

The method of finding out the parameters Θ that maximizes the $p(X|\Theta)$ is called maximum likelihood estimation. If we make function both sides of the logarithm and switch the product into sum, we can get

$$\log p(X|\Theta) = \sum_{i=1}^{N} \log \sum_{j=1}^{M} a_j p_j (x_i|\theta_j)$$

To maximize the value of the above formula, we can introduce Jensen's inequality: for any given convex function $f(x)$, arbitrary discrete probability distribution π_1, $\pi_2, \ldots, \pi_k$ on a set of numbers $a_1, a_2, \ldots, a_k$, then

$$f\left(\sum_{k=1}^{K} \pi_k a_k \right) \geq \sum_{k=1}^{K} \pi_k f(a_k)$$

Further, for any given convex function $f(x)$, $\pi_1, \pi_2, \ldots, \pi_k$ is any arbitrary discrete probability distribution,

$$f\left(\sum_{k=1}^{K} a_k \right) = f\left(\sum_{k=1}^{K} a_k \frac{\pi_k}{\pi_k} \right) \geq \sum_{k=1}^{K} \pi_k f\left(\frac{a_k}{\pi_k} \right)$$

To simplify the discussion, we define

$$\beta_j(x) = \frac{a_j p_j(x|\theta_j)}{\sum_{j=1}^{M} a_j p_j(x|\theta_j)}, \qquad \left(\sum_{j=1}^{M} \beta_j(x) = 1 \right)$$

According to Jensen's inequality, there is

$$\log p(X|\Theta) = \sum_{i=1}^{N} \log \sum_{j=1}^{M} a_j p_j (x_i|\theta_j)$$

$$\geq \sum_{i=1}^{N} \sum_{j=1}^{M} \beta_j(x_i) \log \frac{a_j p_j(x_i|\theta_j)}{\beta_j(x_i)}$$

$$= b(\Theta)$$

$$= \sum_{i=1}^{N} \sum_{j=1}^{M} \beta_j(x_i) \left[\log a_j p_j(x_i|\theta_j) - \log \beta_j(x_i) \right]$$

The above problem is now switched into maximization of the lower bound $b(\Theta)$. Because each Gaussian function $\theta_j = (\mu_j, \sigma_j)$ has two parameters $\theta_j = (\mu_j, \sigma_j)$, when $b(\Theta)$ seeks the maximum, it is to differentiate μ_j and σ_j. Because $\sum_{j=1}^{M} a_j = 1$, hence a_j can be solved by the Lagrangian method:

$$\mu_j = \frac{\sum_{i=1}^{N} \beta_j(x_i) x_i}{\sum_{i=1}^{N} \beta_j(x_i)}$$

$$\sigma_j^2 = \frac{\sum_{i=1}^{N} \beta_j(x_i)(x_i - \mu_j)^T (x_i - \mu_j)}{\sum_{i=1}^{N} \beta_j(x_i)}$$

$$a_j = \frac{1}{N} \sum_{i=1}^{N} \beta_j(x_i)$$

The above three equations can be used as the basis for an iterative solution. According to given data, we can set an initial set of parameter values θ, calculate new parameters $\hat{\theta}$, and execute iteratively. When parameters do not change significantly, the iteration $|\theta - \hat{\theta}|$ is terminated when error is less than the last iteration ε.

3.2.2 Gaussian Transform Algorithm

According to the above GT parameter estimation method, the GT algorithm can be obtained as follows:

GT algorithm
Input: The data sample set $X\{x_i | i = 1, 2, ..., N\}$, the number of Gaussian components M, error ε.
Output: M Gaussian distributions (μ_k, σ_k) and amplitude a_k, $k = 1, ..., M$.
Steps:
   1. Calculating the frequency distribution of the data sample set $X\{x_i | i = 1, 2, ..., N\}$

$$h(y_j) = p(x_i), \quad i = 1, 2, ..., N; j = 1, 2, ..., N'$$

where y is the sample domain space.
   2. Setting initial values of M Gaussian distributions, the kth ($k = 1, ..., M$) initial Gaussian distribution parameters set is

$$\mu_k = \frac{k \times \max(X)}{M+1}, \quad \sigma_k = \max(X), \quad a_k = \frac{1}{M}$$

3. Defining and calculating the objective function

$$J(\theta) = \sum_{i=1}^{N'} \left\{ h(y_i) \times \ln \sum_{k=1}^{M} \left[a_k g\left(y_i; \mu_k, \sigma_k^2\right) \right] \right\}$$

where

$$g\left(y_i; \mu_k, \sigma_k^2\right) = \frac{1}{\sqrt{2\pi}\sigma_k} e^{-\frac{(y_i - \mu_k)^2}{2\sigma_k^2}}$$

4. For the kth ($k = 1, \ldots, M$) Gaussian distribution, according to maximum likelihood estimation, calculating the new parameters of this Gaussian distribution

$$\mu_k = \frac{\sum_{i=1}^{N} L_k(x_i) x_i}{\sum_{i=1}^{N} L_k(x_i)}$$

$$\sigma_k^2 = \frac{\sum_{i=1}^{N} L_k(x_i)(x_i - \mu_k)^T (x_i - \mu_k)}{\sum_{i=1}^{N} L_k(x_i)}$$

$$a_k = \frac{1}{N} \sum_{i=1}^{N} L_k(x_i)$$

where

$$L_k(x_i) = \frac{a_k g\left(x_i; \mu_k, \sigma_k^2\right)}{\sum_{n=1}^{M} \left(a_n g\left(x_i; \mu_n, \sigma_n^2\right)\right)}$$

5. Calculating the estimated value of the objective function

$$J(\theta) = \sum_{i=1}^{N'} \left[h(y_i) \times \ln \sum_{k=1}^{M} \left[a_k g\left(y_i; \mu_k, \sigma_k^2\right) \right] \right]$$

6. Determining the difference between the estimated value of the objective function and the original objective function value, if

$$\left| J(\hat{\theta}) - J(\theta) \right| < \varepsilon$$

output the current parameter estimation; otherwise, skip to step (3).

Next, the age distribution of academicians in the Chinese Academy of Engineering will serve as an example to explain how to use the GT to classify the academicians by age groups.

According to the data on academician ages published on the Chinese Academy of Engineering's website (www.cae.cn), there were a total of 776 academicians aged between 43 and 99 years in April 2012, including 740 males and 36 females; 206 were senior academicians over 80. If we take 1 year as the interval, the frequency distribution of academicians' ages is shown in Figure 3.5.

If we set the Gaussian distribution number $M = 5$ and the iteration termination error $\varepsilon = 0.001$, then the results of GT are shown in Figure 3.6, where the zigzag solid line represents the original distribution, the dotted line is five generated Gaussian distributions, the smooth solid line represents the fitted curve, and the bottom zigzag solid line indicates the absolute errors.

As a result, the 776 academicians are divided into five categories, as listed in Table 3.1: the first category averages 53.1 years old, accounting for 11% of the total; the age of the second category averages 67.0 years old, accounting for 12% of the total; the third category averages 74.2 years old, accounting for 31% of the total; the age of the fourth category averages 77.5 years old, accounting for 31% of the total; the fifth category averages at 83.5 years old, accounting for 15% of the total.

As can be seen from the above results, these five Gaussian distributions overlap quite substantially with each other and the division is confused. Normally, the more initial numbers given for Gaussian distribution, the lower the curve fitting error will be, but there will be more overlap between different categories, and sometimes it is

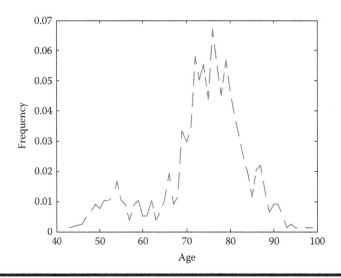

Figure 3.5 The age distribution of academicians in the Chinese Academy of Engineering in 2012.

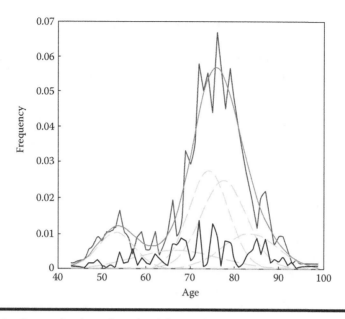

Figure 3.6 Categorization of academicians using GT.

Table 3.1 Five Age Groups of Academicians Divided by GT

Categories	Expected Value	Standard Deviation	Ratio (%)
First category	53.1	4.0	11
Second category	67.0	9.1	12
Third category	74.2	4.4	31
Fourth category	77.5	5.1	31
Fifth category	83.5	6.0	15

difficult to distinguish. So, the difficult problems in GT are how to determine the number of Gaussian distributions, and how to set the initial value of each parameter.

3.3 Gaussian Cloud Transformation

When humans think, they can abstract, analyze, and reason repeatedly at different levels and granularities and can switch naturally between concepts at different granularities. The difficult problem in granular computing research is how to simulate the human adaptability for realizing a concept's classification and clustering.

GT provides a discrete method of continuously attributing data distribution functions, offering a guideline for concept classification and clustering of variable granularity. But GT is only a mathematical fitting process, and it neither considers adaptive cognition under specific situations and themes, nor does it involve semantics. There are some problems in using Gaussian Transform for concept classification.

Firstly, the concept of human cognition is abstract and flexible. An inherent attribute of a concept is being both this and that at the same time. But the hard division of concept by Gaussian transformation determines which concept the element belongs to through the probability value of each element corresponding to each Gaussian distribution. The relationship between element and concept is determined as either this or that, so it cannot reflect "both this and that" and cross the boundaries of adjacent concepts.

Secondly, a Gaussian transform requires the number of concepts that need to be given in advance. If the specified number is too small, it cannot meet the error requirement of data fitting; if the number is too big, then the multiple Gaussian distributions generated will overlap, which is contrary to the general principle of "relationship within the class is strong, and the relationship between the classes is weak," and will lead to confused concepts. Moreover, it is difficult to simulate the variable granularity thinking ability of humans to freely switch between different concept granularities and levels with Gaussian transform.

The cloud model is a dual cognitive model for qualitative–quantitative transformation. A Gaussian cloud has universality in conceptual representation; the contribution of the Gaussian cloud drop group to the concept can be quantitatively calculated. Therefore, we can build a Gaussian cloud transformation to convert any given set of data samples to multiple qualitative concepts that can reflect the uncertainty of the concept boundaries. More importantly, by calculating the ratio of En and He, we can optimize the quantity, granularity, and level of concept so as to prevent a confusion of concepts.

3.3.1 From Gaussian Transformation to Gaussian Cloud Transformation

GT does not involve any semantic meaning of concepts, which raises several questions. How to convert the results to qualitative concepts and measure their clarity? How to reflect the general principles of classification? How to optimize the quantity, granularity, and level of concept? How to show the uncertainty of cognitive levels and the granularities of the concept?

For this purpose, we can use the concept of confusion degree (CD), as defined in Chapter 2, to measure the consensus extent of the concept, and the clarity of the concept division. When $He = 0$, a Gaussian cloud becomes a Gaussian distribution, CD is 0, and concept forms the greatest consensus; with the increase of He, the CD increases, consensus becomes low; when He is equal to $En/3$, CD is equal to 1, the extent of the concept diverges from the cloud into a fog, and it is difficult to form a consensus.

In GT, the consensus degree of the concept is related to the overlap of concepts. The lower the overlap, the clearer the concept, and CD decreases; on the other hand, the more the overlap, indicating that the extension of the concept is more diverged, the more difficult it is to form a consensus, and CD increases. Based on a GT and cloud model, we can construct a GCT to transfer the Gaussian distribution to a Gaussian cloud. According to the law of data classification, closer relations often exist between elements in the same class, and elements in different classes have looser relations. So, we try to use the overlap extent of a Gaussian distribution with its neighbors to compute *En* and *He* of its relevant concept.

In a Gaussian distribution $G(\mu_k, \sigma_k)$, if its weak peripheral elements region is separated from that in other Gaussian distributions, its relevant concept parameter is $En_k = \sigma_k$, $He_k = 0$. Otherwise, their standard variance is set at the same zoom to guarantee that their weak peripheral regions do not overlap. Then a two-zoom parameter α_1, α_2 can be obtained by the formula

$$\mu_{k-1} + 3a_1\sigma_{k-1} = \mu_k - 3a_1\sigma_k$$

and

$$\mu_k + 3a_2\sigma_k = \mu_{k+1} - 3a_2\sigma_{k+1}$$

α_1 is the zoom rate of standard variance where the weak peripheral element region of this concept does not overlap with its left concept. α_2 is the zoom rate of standard variance where the weak peripheral element region of this concept does not overlap with its right concept.

The standard variance range of $G(\mu_k, \sigma_k)$ is $[\alpha \times \sigma_k, \sigma_k]$ and $\alpha = \min(\alpha_1, \alpha_2)$. According to the definition of a Gaussian cloud, *En* is the expectation of standard variance, and *He* is the standard variance of standard variance. Then Gaussian cloud $C(Ex_k, En_k, He_k)$ parameters can be calculated as follows:

$$Ex_k = \mu_k$$
$$En_k = \sigma_k - (1-\alpha) \times \frac{\sigma_k}{2}$$
$$He_k = (1-\alpha) \times \frac{\sigma_k}{6}$$

The CD of the concept is

$$CD_k = \frac{3 \times He_k}{En_k} = \frac{1-\alpha}{1+\alpha}$$

As can be seen, the greater the confusion degree of a concept, the more discrete the extension of that concept will be, and the more it will overlap with adjacent

Table 3.2 GT Division Results Measured by Confusion Degree

Degree of Overlap between Adjacent Conceptual Elements	Standard Deviation Zoom Ratio	Confusion Degree	The Concept Explanation
Same *Ex*	0	1	Atomized
Backbone area overlap	(0, 0.223)	(0.6354, 1)	Confusion
No overlap in backbone area; basic area overlap	(0.223, 0.333)	(0.5004, 0.6354]	Not clear
Peripheral area overlap	(0.3333, 0.667)	(0.2, 0.5004]	A little clear
Weak peripheral area overlap	(0.667, 1)	(0, 0.2]	Clear
No overlap in weak peripheral area	1	0	Very clear

concepts. As a result, it is difficult to form a consensus for that concept. On the other hand, if the confusion degree of the concept is small, the extension of the concept will converge and overlap less with adjacent concepts. So the concept forms a consensus more easily. If the weak peripheral region of the concept (i.e. from 2σ to 3σ), does not overlap other concepts, it indicates a clear concept, its *En* is the standard deviation of the Gaussian distribution, its *He* is 0. In extreme cases, if the expectation of one concept is the same as that of an adjacent concept, then $\alpha = 0$, CD is 1, the concept cannot form a consensus.

The overlap degree of conceptual extension elements can be divided into the backbone area overlap, the basic area overlap, the peripheral area overlap, and the weak peripheral area overlap—as listed in Table 3.2.

3.3.2 *Heuristic Gaussian Cloud Transformation*

The use of GT for concept clustering as mentioned above does not reflect the uncertainty at the edge of concepts, while *En* and *He* of Gaussian cloud not only address the problem, but they can also be used to measure the CD in GT. To this end, the heuristic Gaussian cloud transformation (H-GCT) algorithm is introduced [3].

H-GCT uses a priori knowledge to pre-specify the number *M* of concepts, then applies GT to obtain expectations, standard deviations, and amplitudes of the *M* Gaussian distributions. The *M* expectations of Gaussian distributions are the *M* of the Gaussian cloud. Then, depending on the degree of overlap between the Gaussian distributions, *En*, *He*, and CD are computed for each Gaussian cloud. Thus, the frequency distribution functions of quantitative data are converted into qualitative cognitive concepts; according to the CD, the clarity of concept can be arranged in order.

Heuristic Gaussian Cloud Transformation Algorithm
Input: The data sample set $X\{x_i | i = 1, 2, \ldots, N\}$, the number of concept M.
Output: M Gaussian clouds $C(Ex_k, En_k, He_k) | k = 1, \ldots, M$.
Steps:

1. Calculating the frequency distribution $p(x_i)$ of the data sample set $X\{x_i | i = 1, 2, \ldots, N\}$, let the peak number M of $p(x_i)$ be the initial value; using GT to transform it into M Gaussian distribution

$$G\left(\mu_k, \sigma_k\right) | k = 1, \ldots, M$$

2. For the kth Gaussian distribution, calculating the zoom rate α_k of the standard deviation, the Gaussian cloud parameters of the kth concept are

$$Ex_k = \mu_k$$

$$En_k = \left(1 + \alpha_k\right) \times \frac{\sigma_k}{2}$$

$$He_k = \left(1 - \alpha_k\right) \times \frac{\sigma_k}{6}$$

$$CD_k = \frac{1 - \alpha_k}{1 + \alpha_k}$$

Let us still take the CAE academicians as an example. Adults can generally be divided into five categories by age, as very young, young, middle-aged, elderly, and longevity. We can designate five concepts and H-GCT first uses GT to fit the age curve into five Gaussian distributions; then, according to the degree of overlap between the concepts, it calculates the digital characteristics and CD of each concept. Figure 3.7 shows the results of H-GCT, where the jagged solid line is the frequency distribution of the age, the green cloud drops reflect the five categories, and the black dashed curve is the expectation of Gaussian clouds, the red smooth solid line is the fitting curve of GCT.

As listed in Table 3.3 in the five concepts generated by H-GCT, the Ex of very young age is 53.1 years old; according to the relationship between the CD and the overlap listed in Table 3.2, we can see that the basic area of this concept is independent, overlap only exists between the peripheral area and the young ones and it is a relatively independent category. Backbone area overlap exists with the young, middle-aged, elderly, and longevity overlapping, so these four concepts are confused.

The biggest difference between the GCT and GT is the membership degree of cloud drops, which both guarantees the certainty of the core samples and reflects the "this and that" nature of the samples in the boundary area, so as to achieve a

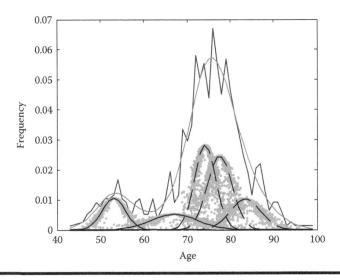

Figure 3.7 Five age groups of academicians divided by H-GCT.

Table 3.3 The Digital Characteristics of the Five Age Groups Divided by H-GCT

The Concept	Ex	En	He	CD	Ratio (%)
Very young	53.1	2.7	0.42	0.468	11
Young	67.0	5.4	1.3	0.723	12
Middle-aged	74.2	2.4	0.64	0.8	31
Elderly	77.5	2.8	0.75	0.8	31
Longevity	83.5	3.4	0.89	0.785	15

soft division of the concepts. For example, in the GCT, a 59-year-old academician can belong to both the very young and the young categories. But in the GT results, because its probability is greater in the young category than in the very young, the former is where it belongs.

We also can divide the age of academicians into young, middle-aged, and elderly from the perspective of a larger granularity. Figure 3.8 shows the three Gaussian clouds generated by H-GCT, and their fitting curves. The jagged solid line is the original frequency distribution, the cloud drops reflect the three Gaussian cloud concepts, the dashed line is the expectation curve, and the smooth solid line is the fitting curve of GCT.

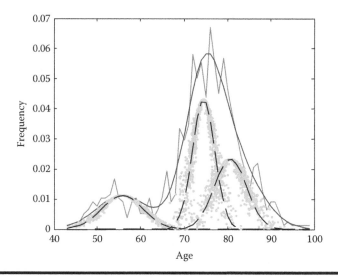

Figure 3.8 Three age groups of academicians divided by H-GCT.

Table 3.4 The Digital Characteristics of the Three Age Groups Divided by H-GCT

The Concept	Ex	En	He	CD	Ratio (%)
Young	55.8	4.8	0.39	0.244	16.6
Middle-aged	74.5	2.6	0.58	0.675	46.9
Elderly	80.6	3.7	0.84	0.675	36.5

As listed in Table 3.4, in the three concepts generated by H-GCT the *Ex* for young is 55.8 years old. According to the relationship between the CD and overlap listed in Table 3.2, we can see that there is overlap with middle-aged in a small part of the peripheral area, so the division is relatively clear and is a more recognizable category. There is overlap in the backbone area between middle-aged and elderly, so they are confusedly divided concepts.

As can be seen from these two division results, due to a serious conceptual overlap, both the five and the three concepts for dividing the academicians by age might not comply with human cognition.

H-GCT provides a soft division method that can transform quantitative data to multiple qualitative concepts, thus addressing the "this and that" problem in the boundaries. But the H-GCT method does not offer a solution for the problem of confused concept division, so the problems of optimizing the number, granularity, and hierarchy of concepts still exist.

3.3.3 Adaptive Gaussian Cloud Transformation

To automatically find the appropriate number of concepts with little overlap, an adaptive Gaussian cloud transformation (A-GCT) algorithm [3] has been proposed.

A-GCT does not pre-specify the number of concepts. Instead, it starts from the statistical distribution of actual data samples and automatically forms a number of concepts that are suitable for human cognition. The A-GCT process reflects the extraction and clustering process of human cognition from low level and fine-grained concepts to those that are high level and coarse-grained, hence it can achieve variable granular computing.

Common sense tells us that compared to the data values that appear with low frequency, data values that appear with high frequency make a greater contribution to a qualitative concept. So we can count the number of peaks in the frequency distribution of the data sample and take it as the initial conceptual number of the Gaussian cloud transformation. Then we can use H-GCT to generate M concepts according to the confusion degree of concept to make a Gaussian cloud transformation strategy. For example, if each CD ≤ 0.5004, it means there is no basic area overlap between this concept and its adjacent concepts. Iterative convergence is carried out by the continuous use of H-GCT, and finally concepts that meet the requirements of iterative termination conditions are formed.

Adaptive Gaussian Cloud Transformation (A-GCT)
Input: The data sample set $X\{x_i|i = 1, 2, \ldots, N\}$, confusion degree threshold β.
Output: M Gaussian clouds $C(Ex_k, En_k, He_k)|k = 1, \ldots, M$.
Steps:
1. Let the number of peaks m of the frequency distribution $p(x_i)$ of the data sample set $\{\langle x_i, y_i, z_i \rangle|i = 1, 2, \ldots, N\}$ be the initial value of the concept number.
2. Use the H-GCT algorithm to cluster the data set $X\{x_i|i = 1, 2, \ldots, N\}$ into M Gaussian clouds

$$C\left(Ex_k, En_k, He_k \right)|k = 1, \ldots, M$$

3. Judge CD of each concept, if $CD_k > \beta$, $k = 1, \ldots, M$, then the concept of the number $M = M - 1$, jump to step 2; otherwise, output M concepts which $CD_k < \beta$:

$$C\left(Ex_k, En_k, He_k \right)|k = 1, \ldots, M$$

This algorithm can be discussed further. If there are too many initial concepts, the number of iterations of H-GCT will be increased, and the algorithm will be more complex. During actual application, the data frequency distribution is discrete, but not a continuous probability density distribution function and serious fluctuations may exist. For example, in the age distribution of CAE academicians, as shown in Figure 3.5, which has 21 peak values, it is better to smooth the peak to reduce the fluctuation of the frequency curve. As can be seen from the algorithm in step 3, in the result of each transformation, as long as there is a CD > β, the number of concepts will be decremented by 1. H-GCT is repeatedly used until the termination condition is satisfied. Under normal circumstances, the results of A-GCT are independent from the initial number of concepts.

In Figure 3.6, the academicians are divided into five categories. There is overlap in the backbone area of the young, middle-aged, elderly, and longevity. We might also start from the five concepts to analyze the results of the Gaussian cloud transformation and iterative convergence by A-GCT. First, we reduce one concept to generate concepts as shown in Figure 3.9. The cloud drops reflect the four Gaussian cloud concepts; the dashed line is the expectation curve, and the smooth solid line is the fitting curve.

So, as listed in Table 3.5, there is overlap in the backbone area of the middle-aged, elderly, and longevity, and the confusion degrees of the concept are high. Then we reduce one more concept and use H-GCT to generate three concepts, as shown in Table 3.4, in which there is overlap in the backbone area of the middle-aged and the elderly, and the confusion degrees of concept are high. Finally, we further reduce another concept and use H-GCT, then two concepts are generated, as shown in Figure 3.10.

As listed in Table 3.6, A-GCT generates two concepts, young and elderly, with no overlap in their basic or peripheral regions and only partial overlap in the weak peripheral zone. *Ex* of young academicians is 53.0 years old, *En* is 3.3 years old, they

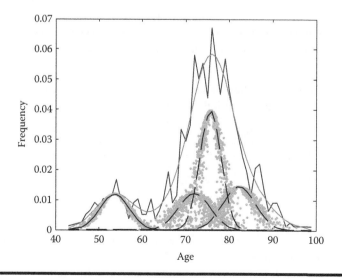

Figure 3.9   Four age groups of academicians divided by A-GCT.

Table 3.5   The Digital Characteristics of the Four Age Groups Divided by A-GCT

The Concept	Ex	En	He	CD	Ratio (%)
Young	53.6	3.4	0.32	0.285	12.98
Middle-aged	72.1	3.7	0.99	0.7998	20.40
Elderly	75.8	2.5	0.66	0.7998	44.06
Longevity	82.3	3.5	0.92	0.6604	22.56

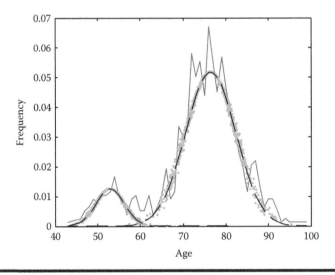

Figure 3.10   Two age groups of academicians divided by A-GCT.

Table 3.6   The Digital Characteristics of the Two Age Groups Divided by A-GCT

The Concept	Ex	En	He	CD	Ratio (%)
Young	53.0	3.3	0.16	0.145	14
Elderly	76.4	5.9	0.29	0.145	86

form 14% of the total; *Ex* of elderly academicians is 76.4 years old, *En* is 5.9 years old, making up 86% of the total. Both have a low degree of confusion and are more recognizable mature concepts.

Therefore, using A-GCT to cluster the academicians by age finally leads to two concepts more compliant with human cognition. In particular, academicians around

60 years old are at the junction of these two concepts, which reflects the objective situation that there were fewer academicians in the 1950s.

A-GCT is more suitable for clustering larger amounts of data. The statistical distribution curve of large data samples is usually more stable. ArnetMiner (www. arnetminer.com) is a website developed by the Software and Knowledge Engineering Laboratory, Tsinghua University, specifically for the study of social network mining and academic research. From 2006 to March 26, 2012, there were 988,645 registered users from 196 countries. The users' age distribution is shown in Figure 3.11.

If we use A-GCT to cluster the ArnetMiner users by age and do not allow overlap in the basic region, we can generate three concepts, as shown in Figure 3.12. The jagged solid line is the original frequency distribution, the cloud drops reflect three Gaussian cloud concepts, the dashed line is the expectation curve, and the smooth solid line is the fitting curve of GCT. As listed in Table 3.7, ages are clustered into three concepts: the first is young, 30.4 years old, whose granularity is 2.1 years old, accounting for 65%; the second is middle-aged, 40.9 years old, whose granularity is 3.8 years old, accounting for 30.1%; the third concept is the elderly, 57 years old, whose granularity is 5.7 years old, accounting for 4.9%. Their basic areas do not overlap, showing a clear concept division that complies with human cognition.

A-GCT can simulate human cognition to realize automatic data clustering based on the concept of CD limitation strategy. When it is demonstrated that Gaussian clouds are generated every time in the A-GCT process, we can simulate the variable granular ability of human cognition. Figure 3.13 shows the variable granular cognitive process from five to two concepts, using A-GCT for clustering academicians by age. Figure 3.14 shows a pan-tree for academician ages built by A-GCT.

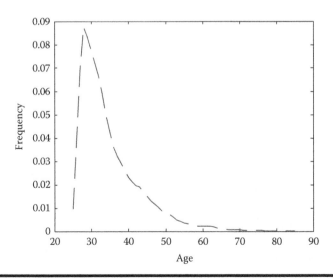

Figure 3.11   Distribution curve of ArnetMiner users' ages.

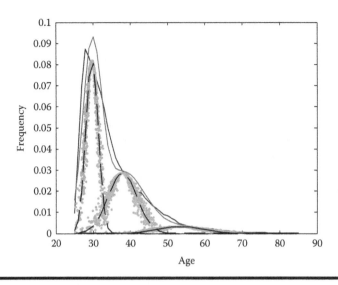

Figure 3.12   Three groups of ArnetMiner users divided by A-GCT.

Table 3.7   The Digital Characteristics of the Three Age Groups of ArnetMiner Users

The Concept	Ex	En	He	CD	Ratio (%)
Young	30.4	2.1	0.29	0.408	65
Middle-aged	40.9	3.8	0.55	0.437	30.1
Elderly	57.0	5.7	0.83	0.437	4.9

3.3.4 High-Dimensional Gaussian Cloud Transformation

Theoretically, the GCT algorithm can also be used to process two-dimensional or high-dimensional attribute data. For two-dimensional attribute data, the overlap degree of the two concepts is calculated by overlapping regions of the two ellipses. For three-dimensional attribute data, the overlap degree of the two concepts is calculated by overlapping regions of the two ellipsoids. Attribute data over three dimensions cannot be visualized in the coordinate space. The overlap degree between the concepts can be calculated separately by the projection of the concept on each dimension. As a result, the CD of the concept is a multidimensional vector.

To simplify the problem, in this section, the projections of each concept on each dimension are set to be the same. The conceptual extension of the two-dimensional attribute data is a circular area with an uncertain radius, where the center (Ex_1, Ex_2) is the expectation, the radius is a random number with triple En as the expectation, and He as the standard deviation. The conceptual extension of the three-dimensional

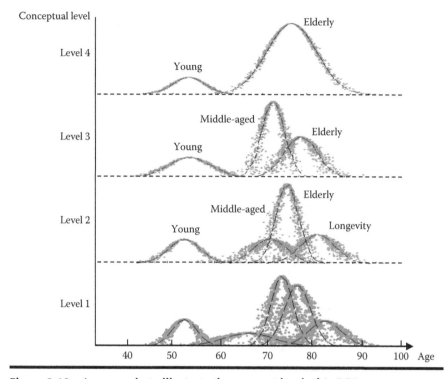

Figure 3.13 An example to illustrate the concept level of A-GCT.

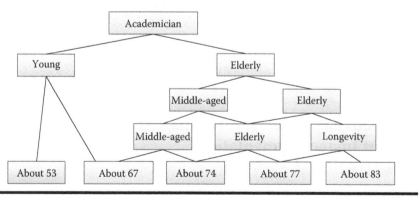

Figure 3.14 Pan-tree constructed by A-GCT.

attribute data is a spherical ball with an uncertain sphere diameter, where the center of the sphere (Ex_1, Ex_2, Ex_3) is expectation, the sphere diameter is a random number with triple En as the expection, and He as the standard deviation. The CD of multidimension concepts is calculated at the same as that for one-dimensional attributes, namely, through the relationship between the distance of expectations of two concepts and their radii.

The following algorithm is given for a three-dimensional GCT with the same *En* and *He* in each dimension attribute.

Three-dimensional GCT algorithm
Input: Data sample set $\{\langle x_i, y_i, z_i \rangle | i = 1, 2, \ldots, N\}$, Confusion Degree threshold β.
Output: M concept $C(\langle Ex_k, Ey_k, Ez_k \rangle, En_k, He_k)$, $k = 1, \ldots, M$
Steps:

1. Let the number of peaks M of the frequency distribution of the data sample set $\{\langle x_i, y_i, z_i \rangle | i = 1, 2, \ldots, N\}$, as the initial value of the concept number
2. Use three-dimensional Gaussian transformation to cluster the data into M components

$$G\left(\langle \mu x_k, \mu y_k, \mu z_k \rangle, \sigma_k\right), \quad k = 1, \ldots, M$$

3. Generate m concepts and calculate CD of each concept

$$C\left(\langle Ex_k, Ey_k, Ez_k \rangle, En_k, He_k\right), \quad k = 1, \ldots, M$$

4. Judge CD of each concept, if $CD_k > \beta$, $k = 1, \ldots, M$, then the concept of the number $M = M - 1$, jump to step 2; otherwise, output M concepts with each $CD_k < \beta$:

$$C\left(\langle Ex_k, Ey_k, Ez_k \rangle, En_k, He_k\right), \quad k = 1, \ldots, M$$

Based on the GT, GCT uses *He* to solve the soft division problem of the uncertainty area between concepts and the CD to measure the concept consensus; then, by using the CD threshold, GCT realizes the generation, selection, and optimization of a concept's number, granularity, and level. In this sense, GCT is not only a clustering process, but is also a process of granular computing.

3.4 Gaussian Cloud Transformation for Image Segmentation

Just like the Fourier transform, Gaussian cloud transformation gives a general cognition tool that not only converts data collection to concepts at different granularities, but also realizes a switch between them. As it can solve various granular computing problems, it can be used in a broad range of applications. Here are a few typical cases.

3.4.1 Detection of the Transition Zone in Images

With the development of computer graphics, simple image segmentation has been unable to meet individual needs. Sometimes what really interests people is the transition zone in an image. It has always been a difficult problem to simulate the cognitive ability of natural human vision to complete image segmentation. Gaussian cloud transformation, as a method of simulating the ability of human cognition in variable granular computing, has the advantage of being able to process uncertain information. Therefore, finding the uncertain region in an image is an important ability of Gaussian cloud transformation.

Laser cladding is a new surface modification technology. It adds cladding materials to a substrate surface and fuses them with a thin layer using a laser beam, thus forming a combination surface for metallurgical cladding. Laser cladding technology has broad industrial applications for which reliable and effective feedback is the key, namely, how to obtain an accurate laser height. If only the gray value attribute of an image is considered, then an image can be seen as a numerical matrix in the interval [0, 255]. By counting the number of pixels corresponding to each gray value, we can draw up a gray histogram of the image, that is, the frequency distribution of gray level pixels. Figure 3.15 shows a 256 × 256 pixel, the laser cladding image with gray value of [0, 255] and its gray histogram. The white area is high energy density laser and the black area is the background color; meanwhile, there is an important transition zone between foreground and background.

If we take the gray value of the image pixel as data collection and do not allow overlap in the basic region of the concept, an adaptive Gaussian cloud transformation will ultimately generate three concepts, as shown in Figure 3.16a. The image is divided into three regions according to the three generated concepts, as shown in Figure 3.16b.

As listed in Table 3.8, the laser cladding image is clustered into three concepts by using A-GCT: the first is with a black background whose expectation of gray is about 101.49; the second is the gray transition zone whose expectation of gray is about 191.38; the third is the white laser area whose expectation of gray is about 253.23. The white area is the clearest concept with a minimum CD of 0.246.

In this laser cladding image, the gray transition zone between black background and white laser area has an obvious gray feature, forming a new concept with an independent connotation and extension. This has not been recognized in many other image segmentation algorithms, which roughly divide the image into the two categories of black background and white laser area, while constantly looking for fuzzy boundaries in the transition zone. The results of classic image segmentation methods, such as C-means, Otsu, and fuzzy C-means, are shown in Figure 3.17. Although the images are all divided into two parts, black and white, their boundary lines vary. The results of a C-means segmentation method is near the black background, and those of Otsu segmentation and fuzzy C-means are close to the white laser area.

The transition zone can be clearly detected by GCT and its discovery and extraction is the focus of image segmentation in laser cladding image.

Of course, the GT is also a commonly used method in image segmentation. If the concept number is specified as three, the transition zone can also be obtained. Compared to the GT, GCT provides a soft target edge segmentation method.

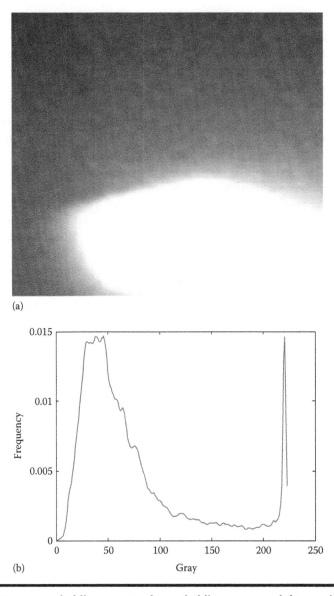

(a)

(b)

Figure 3.15   Laser cladding map: (a) laser cladding map 1 and (b) gray histogram.

The membership degree of each pixel to each concept is obtained by using a forward Gaussian cloud algorithm. Comparing all the membership degrees of each pixel, we can judge which concept a pixel belongs to. These membership degrees are random, with a stable tendency, which can both ensure the concept's core region is divided correctly, and depict the uncertain boundaries between adjacent concepts. Figure 3.18 shows the uncertain boundaries of the three concepts divided

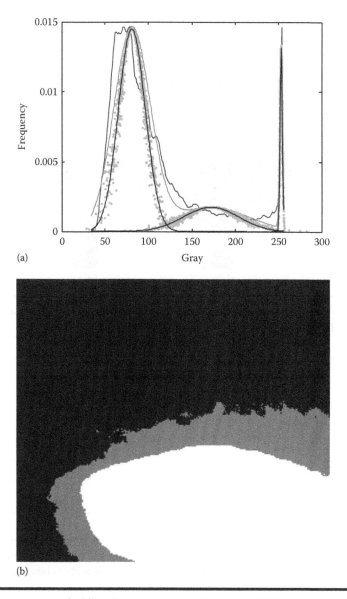

(a)

(b)

Figure 3.16 Laser cladding image segmentation by A-GCT: (a) result of A-GCT and (b) three regions divided by A-GCT.

by Figure 3.16. This is impossible to implement with conventional GT, since GT determines which concept the pixel belongs to based on the probability values of each pixel corresponding to each Gaussian distribution. The uncertainty between two concepts' boundaries cannot be reflected, so the junction of the two regions often shows an unnatural sawtooth.

Table 3.8 The Digital Characteristics of the Three Regions

The Concept	Ex	En	He	CD	Ratio (%)
Black background area	101.49	17.6	1.75	0.3	80.08
Gray transition zone	191.38	25.1	2.48	0.3	15.52
White laser area	253.23	1.3	0.11	0.246	4.39

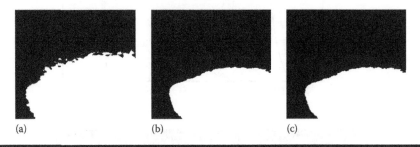

(a) (b) (c)

Figure 3.17 Segmentation results using C-means, Otsu, and fuzzy C-means: (a) C-means, (b) Otsu, and (c) fuzzy C-means.

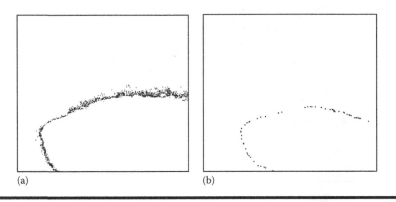

(a) (b)

Figure 3.18 Extraction of the uncertainty edges in laser cladding image 1: (a) boundaries of black background and the transition zone and (b) boundaries of the transition zone and the laser area.

Background, foreground, transition zone, and the uncertainty boundaries of the target are fully reflected in the laser cladding image. GCT uses the statistical characteristics of grays, adaptively extracts multiple concepts with different granularities, finds the transition zone, and clearly presents the uncertain boundaries of the transition zone. Figure 3.19 shows another typical laser cladding segmentation result using Gaussian cloud transformation.

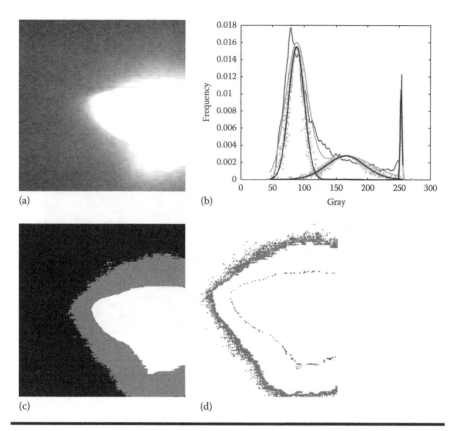

Figure 3.19 Image segmentation by A-GCT in laser cladding image 2: (a) laser cladding image 2, (b) result of A-GCT, (c) result of image segmentation, and (d) extraction of uncertainty boundaries.

3.4.2 Differential Object Extraction in Image

A-GCT not only detects the transition zone in an image, but also extracts clear objects from images based on the concept confusion degree so as to achieve a differential object extraction in the image.

It can be seen from Tables 3.8 and 3.9 that, compared to the black background and gray transition zone, the CD of the white laser area is minimal. The white laser area is a concept that is much closer to maturity. Therefore, according to the principle of difference priority in human cognitive processes, the clearest white laser areas can be extracted from Figures 3.15 and 3.19, with the results shown in Figure 3.20 and the background set to be black.

C-means, Otsu, and fuzzy C-means methods all intend to extract a significant white laser area from the background. Compared to these three segmentation

Table 3.9 The Digital Characteristics of the Three Regions of Laser Cladding Image 2

The Concept	Ex	En	He	CD	Ratio (%)
Black background area	87.97	12.2	1.60	0.393	65.9
Gray transition zone	165.61	30.4	3.98	0.393	29.3
White laser area	253.08	1.5	0.10	0.205	4.8

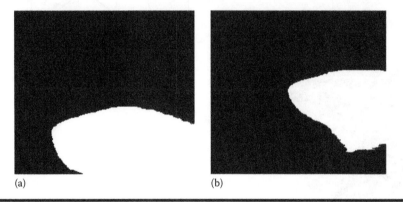

(a) (b)

Figure 3.20 Extraction of clearest goals in laser cladding based on CD: (a) clearest goal in laser cladding image 1 and (b) clearest goal in laser cladding image 2.

methods, what are the results of using GCT for white laser area extraction? Misclassification error (ME) [4] can be used to measure them.

ME refers to the pixel ratio where segmentation results include background pixels incorrectly deemed to be objects, and object pixels incorrectly deemed to be background. It indicates the difference between image segmentation results and human observation.

$$ME = 1 - \frac{|B_0 \cap B_t| + |F_0 \cap F_t|}{|B_0| + |F_0|}$$

where in the reference image, background and objects are denoted as B_0 and F_0, respectively; in the segmented image, background and objects are denoted as B_t and F_t respectively; $|B_0 \cap B_t|$ is a set of pixels that are classified correctly as background; $|F_0 \cap F_t|$ is a set of pixels that are classified correctly as objects. $|\cdot|$ represents the potential of the set. ME values between 0–1, 0 indicates the best segmentation results without misclassification, 1 indicates a complete segmentation error. The smaller the ME is, the higher the quality of the segmentation.

Figure 3.21 are the resultant images by artificial segmentation. They can be seen as reference images to compare the effectiveness of different segmentation algorithms.

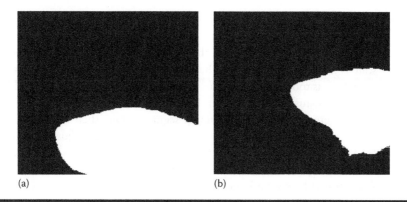

(a)   (b)

Figure 3.21   Reference image by artificial segmentation: (a) reference image 1 and (b) reference image 2.

A quantitative comparison of the four segmentation methods is shown in Table 3.10. The ME of A-GCT is lower than 2%, better than the other three algorithms. But the running time of the A-GCT algorithm is obviously longer than that of other algorithms because there are too many peak values obtained from the gray value distribution, which leads to too many cycles of the algorithm. In the two experiments, the initial peak values were 23 and 19. The reduction process of peak value is the optimization process of concept quantity. When dealing with big data, multi-granularity, and multi-concepts, this disadvantage will turn into an advantage.

Table 3.10   Comparison of the Four Segmentation Methods

Original Graph	Evaluation	C-Means Method	Otsu Method	Fuzzy C-Means Method	A-GCT
Laser cladding image 1	Misclassification number	6584	2209	2447	691
	Misclassification error (%)	12.1	3.9	4.32	1.22
	Running time (s)	1.83	0.016	0.015	4.2
Laser cladding image 2	Misclassification number	13991	5600	5360	861
	Misclassification error (%)	21.3	8.54	8.18	1.31
	Running time (s)	1.70	0.016	0.031	5.8

The above examples show that for those images with significant statistical differences, A-GCT can realize concept extraction and image segmentation according to the principle of non-overlapping in the basic zone. However, if difference in the gray scale is not clear, for example, the image of a snake in the desert shown in Figure 3.22, which is too similar to the image background, and whose overall histogram shows only a Gaussian distribution. In such a case, it will be difficult to form two or more concepts and complete an image segmentation. Then we need to raise more stringent requirements for the confusion degree, such as adjusting β in the interval (0.6354, 1), which allows an overlap in the backbone area of some concepts. When β = 0.66, we can extract five concepts from the image, as listed in Table 3.11.

The black area with expectation of 42.9 has the smallest CD and is the clearest concept in the whole image. If the corresponding pixel area is marked as a black foreground object and others as white background, we get the segmentation result as shown in Figure 3.22d.

This example shows that when there is no obvious difference between the concepts, we can increase the threshold of CD or add to the number of concepts to achieve object segmentation at a finer granularity concept level.

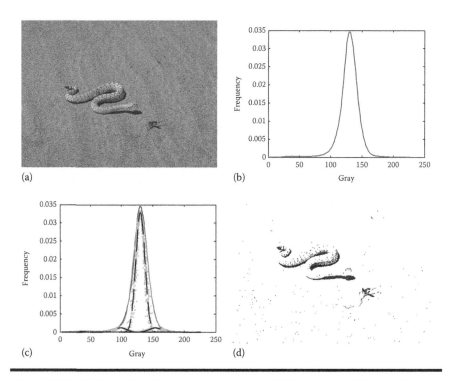

Figure 3.22 Differential objects extraction by A-GCT: (a) original image, (b) Gray histogram, (c) result of A-GCT results, and (d) the clearest area.

Table 3.11   The Digital Characteristics of Five Regions in the Desert Snake Image

The Concept	Ex	En	He	CD	Ratio (%)
Black area	42.9	11.7	0.89	0.22	1.24
Gray area	98.9	10.5	1.62	0.46	4.99
Black gray area	126.6	5.9	1.28	0.65	62.30
White gray area	138.8	5.7	1.24	0.65	30.20
White area	169.8	10.1	1.30	0.39	1.27

As mentioned above, the GCT can be applied to extract concepts of both one- and high-dimensional data, so it can be used for color image segmentation.

Figure 3.23 shows a color image with a resolution of 1024 × 717 pixels. A pair of lovers sits in a background of pink and white, outlined by the sea and light shining through the clouds, forming an artistic picture. If we regard each point in the image as a cloud drop, including three attributes of red, green, and blue, we can use the three-dimensional Gaussian cloud transformation to extract the concept and the differential objects.

GCT automatically generates three color concepts, as listed in Table 3.12, where the hyper entropy of the black character area is 0, and the CD is also 0, indicating that this concept is a mature one, largely different from background, light, and sea level in the image, so it is the clearest object.

The result of segmenting the color image by using three extracted color concepts is shown in Figure 3.24. Because both light and sea are a mix of white and pink, the boundaries of the object area are uncertain. The boundary of the two black

Figure 3.23   A color image.

Table 3.12   Segmentation of a Color Art Image by A-GCT

The Concept	Ex			En	He	CD	Ratio (%)
	Red	Green	Blue				
Black character area	38.66	22.56	4.7	24.6	0	0	8.3
Pink background area	220.18	149.56	80.17	14.3	0.133	0.03	57.3
Intersection of pink and white	252.81	210.46	167.47	23.88	0.223	0.03	34.4

(a)   (b)

Figure 3.24   Color image segmentation by A-GCT: (a) segmentation result and (b) extraction of the clearest goal.

character areas is clear, which can be extracted from the image as a good object. It proves that Gaussian cloud transformation is effective for color image segmentation. More importantly, it embodies the principle of large-scale priority in the process of human cognition.

Humans can quickly and accurately find their area of interest in complex scenes when they perceive the environment visually. Psychological studies suggest that selective attention is one of the important abilities of human natural vision, which is often manifested as a priori knowledge priority, large-scale priority, moving target priority, foreground priority, and difference priority.

Here are two examples to illustrate how to simulate large-scale and difference priority in human visual cognition through GCT. Large-scale priority is when the overall concept is formed through cognition division of the overarching information from the large scale, which then combines the local feature information to classify, recognize, and identify the specific area. Difference priority refers to giving most attention to areas that are largely different from their surroundings in color, shape, brightness, and so on.

The amplitude of concepts can be used to formulate large-scale priority. GCT can measure the clarity degree of object segmentation by CD to formalize difference priority.

Figure 3.25a shows two images taken from the Berkeley standard image database [5]. The top picture is made up of gray sky, black wood, and white moon. The bottom picture is made up of white background, eagle, and tree branches.

Two images are then extracted by using A-GCT.

The top image is clustered into three concepts: the black forest area with a gray value of about 35.7; the sky area with a gray value of about 78.7; and the white moon area with a gray value of about 175.1. As shown in Table 3.13, the gray sky area takes up the largest proportion (66.5%) of the whole image, which is set to be white, with the other areas set to be black. Figure 3.25b shows the segmentation

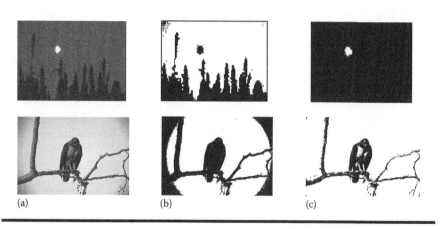

(a) (b) (c)

Figure 3.25 Large-scale priority and difference priority simulated by GCT: (a) original image, (b) large-scale object extraction, and (c) difference object extraction.

Table 3.13 The Digital Characteristics of Three Regions in the Forest Image

The Concept	Ex	En	He	CD	Ratio (%)
Black forest area	35.7	12.27	0.44	0.11	32.9
Gray sky area	78.7	2.89	0.11	0.11	66.5
White moon area	175.1	25.16	0	0	0.6

results where the moon and forest areas are split out from the background, reflecting the large-scale priority; although the third concept of a white moon area only takes up 0.6%, its CD is 0, so it has the clearest division. When it is set to be white and the other areas are set to be black, the segmentation results are shown in Figure 3.25c. The object with the most notable difference is extracted from the background, reflecting the difference priority principle.

GCT combines probability statistics with cognitive concepts. Based on the data's statistical properties and the universality of the Gaussian cloud, it simulates the concept clustering process of different granularities in human cognition. It can effectively realize the formalization of large-scale priority and difference priority principles in natural visual perception by using confusion degree, amplitude, and other parameters. It is more universal and statistical than traditional segmentation methods, so it may be more suitable for big data processing.

References

1. S. Julia, B. Kevin, N. Nikos et al., Developing the next-generation climate system models: Challenges and achievements, *Philosophical Transactions of the Royal Society A—Mathematical Physical and Engineering Sciences*, 367(1890), 815–831, 2009.
2. J. A. Bilmes, A gentle tutorial of the EM algorithm and its application to parameter estimation for Gaussian mixture and hidden Markov models, International Computer Science Institute, Berkeley, CA, April 1997.
3. Y. Liu, Based on cloud the research of granular computing based on cloud, PhD thesis, Tsinghua University, Beijing, China, 2012.
4. A. Z. Arifina and A. Asano, Image segmentation by histogram thresholding using hierarchical cluster analysis, *Pattern Recognition Letters*, 27(13), 1515–1521, 2006.
5. D. Artin and C. Fowlkes, eds., The Berkeley segmentation dataset and benchmark, http://www2.eecs.berkeley.edu/Research/Projects/CS/vision/bsds/.

Chapter 4

Data Fields and Topological Potential

It is a strong human instinct to understand the world. People see the world as a long evolutionary process and have achieved many significant milestones in science, for example, modeling the structure of the atom and the divisibility of matter in physics, the periodic table of chemical elements, the Big Bang theory, continental drift, the theory of evolution, and so on. The famous physicist Tsung-Dao Lee, 1957 Nobel laureate in physics, in his lecture entitled "The Challenge from Physics," said that the essence of physical progress in the twentieth century was simplicity and induction. All sciences, including physics, astronomy, chemistry, and biology, are essentially a systematic approach to new and accurate abstractions of natural phenomena. The simpler the abstractive narration, the more far-reaching the scientific invention, and the more extensive its application. These incisive remarks inspired us to ask a profound question in cognitive science, that is, is there a similarity between human subjective perception and the objective cognition of the world? and is it possible to borrow ideas from objective cognition to understand human self-cognition and promote research on artificial intelligence? Perhaps, it will become a significant aspect of the development of cognitive science in the twenty-first century. In this book we call it "cognitive physics" [1,2].

4.1 Data Field

4.1.1 Using Field to Describe Interactive Objects

Physical understanding and description of the objective world—whether in mechanics, electromagnetism, modern physics, and so on—involve the concepts of interaction and field at different scales and levels. Modern physics even thinks that field is one of the basic forms for existing materials and no particles can exist without a related field.

So far, physicists have found four fundamental forces in the universe: gravitational, electromagnetic, strong interaction, and weak interaction. Newton's law of universal gravitation states that there are mutually interactive gravitational fields and gravitational potential energy in a multi-particle system, and further, that all objects in the universe will converge due to the gravitational attraction between them. According to Coulomb's law, the electric field between charged points at a distance can be visually portrayed with the help of electric field lines and equipotential lines (or surfaces), where the electrical potential has the same value everywhere. Nuclear physics believes that strong interactions are achieved between nucleons, and between nucleons and mesons, through the inter-quarks exchange of gluons. According to the universal Fermi theory, the weak interaction, as a point interaction, is unrelated to any field. In 1984, Carlo Rubbia and Simon Van der Meer were awarded the Nobel Prize for their outstanding contribution in discovering the field quanta of weak interaction.

Although the four fundamental forces exist at different scales and belong to different branches of physics, physicists have been attempting to incorporate the associated theories, which has become an important trend in the research. They have been exploring the possibility of using a unified theory to describe the four types of physical action. Einstein devoted his later life to unifying gravitational and electromagnetic forces, but did not succeed. In the late 1960s, Steven Weinberg and his colleagues introduced the unified electroweak theory to describe electromagnetic effects and weak interactions, which greatly encouraged physicists to produce a grand unified theory (GUT) in the 1970s. According to the GUT, at extremely high energy (almost as high as 10^{24} eV), electromagnetic force, strong interaction force, and weak interaction force can be consistent. The Big Bang theory even points to gravitational force being unified with the other three forces in the very early stage of the universe.

People's own perceptions and thinking processes are essentially a reduction process from data to concept, and then to knowledge. According to cognitive physics, we can use the atomic model to describe this reduction process in human cognition. We can draw on objective cognition for a subjective perception of the world. If we take the physical form of electrons rotating nuclei as a model, we can regard the fundamental nucleus in the cognitive atomic model as expectation, with entropy and hyper entropy as attached to that nucleus. Expectation, entropy, and hyper entropy correspond to a nucleus, and the many sample points around the conceptive extension correspond to a group of electrons rotating the nucleus. This is our conceptual atomic model. We can also learn from other theories in physics to describe subjective cognition. In this chapter we try to extend cognition theory on the objective world in modern physics to cognition of the subjective world, and to introduce an interaction between data objects and the concept of field, so as to establish the cognitive field and formalize the description of the complex association between original, chaotic, and shapeless data objects. In the cognitive field, concepts, linguistic values, words, and even data points in domain space can be regarded as the objects of interaction in the field space.

Knowledge discovery from ultra-large-scale relational databases is one of the frontiers in computer science. Given a logic database composed of N records and with an attribute of M dimensions, the process of knowledge discovery begins with the underlying distribution of N data points in the M-dimensional universal space. If each data object is viewed as a "point charge," or "particle," it will exert a force on other objects in its vicinity. In this way, the interaction of all data points will form a field. The cognitive process from quantitative data, to qualitative concept, to knowledge can be seen as study of interactions and relationships between these objects through the field at different granularities so as to mimic the human cognitive process.

Let us suppose an employee database containing information on 3000 employees' salaries and length of service. A two-dimensional data field is created, as shown in Figure 4.1a, in which the global topology of equipotential lines reveals the clustering properties of the underlying data. We can see that all the data cluster into classes of A, B, C, D, or E. Where class A represents long service time and lower salaries; class B represents long service time and high salaries; class C represents longer service time and higher salaries; class D represents short service time and higher salaries; class E represents short service time and low salaries. Then, classes A and B constitute the same lineage class AB at level 2, indicating long years of service; classes D and E constitute class DE at level 3, indicating short service time, and so on until, finally, all individuals are merged as one class at level 5 to form the largest lineage class ABCDE, as shown in Figure 4.1b. This simple example tells us that we may use the field to describe the interaction between objects.

4.1.2 From Physical Field to Data Field

The concept of field was first proposed by an English physicist, Michael Faraday, in 1837. In his opinion the non-contact interaction between particles, as in gravitational, electrostatic, and magnetic forces, cannot be realized without a certain medium called "field." With the development of field theory, people have abstracted that notion into a mathematical concept.

If every point in space Ω corresponds to the determinate value of a physical variable or mathematical function, the field of such variable or function is said to be identified in space Ω.

Obviously, the field describes the distribution rule of such physical variable or mathematical function in the space. If every point in the field space is associated with a single number or magnitude, the related field is called a "scalar field," for example,

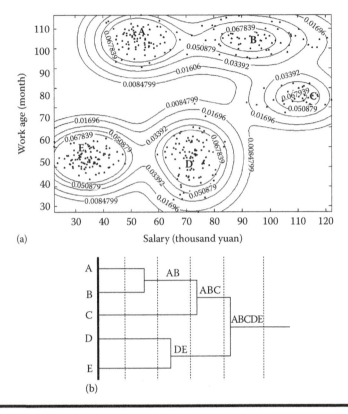

(a)

(b)

Figure 4.1   Data field and class spectrum for 3000 objects formed by salary–service time: (a) equipotential line graph and (b) class spectrum.

temperature, density, or electric potential fields. By contrast, if the field is characterized by vectors closely related to magnitude and direction, it is called a "vector field."

In physics, a vector field is a general concept represented by electric field, magnetic field, gravitational field, velocity field, and so on. According to the various properties or conditions of the material, the vector field is viewed as either a source field, or a rotational field, or a superposition of both. For instance, gravitational, static electric, and static nuclear force fields are source fields; a magnetic field is only a rotational field; and the field produced by electric flux changing with time is both a source and a rotational field.

A source field is the most frequently discussed vector field in fundamental physics, which has been elaborated in some basic theorems, such as Newton's law of universal gravitation and Coulomb's law. Its main characteristics are that every object located in a source field will exert a certain force, and that field lines start from the sources and end at the sinks. As the value of a physical quantity at each point in the field changes with time, the source field can be further divided into a constant source field or a time-varying source field. In a constant source field, independent from time, there must be a scalar potential function φ(*r*) corresponding to a vector intensity function *F*(*r*), to enable them to be linked by a differential operator *F*(*r*) = ∇φ(*r*). Thus, a constant source field is also called a "conservative field" or a "potential field," where potential means the work done to move a unit mass point—such as the unit mass in a gravitational field or unit positive charge in a static electric field—from point A to the reference point. In a potential field, the distribution of potential energy is determined uniquely by the relative position of interactive particles in space. Since the potential energy difference between point A and the reference point is a definite value, the potential function is usually deemed simply as a single-valued function of the coordinates, whether mass points exist or not.

For example, in a gravitational field produced by particles with mass *m*, the potential value in any point is expressed as

$$\varphi(r) = \frac{G \times m}{\|r\|}$$

where

  r is the radial coordinate of any point in the spherical coordinate system, with its origin point *m*
  G is the gravitational constant

If there are *n* mass points in the space, the total potential at point *r* equals the algebraic sum of the potentials produced individually by every mass point,

$$\varphi(r) = G \times \sum_{i=1}^{n} \frac{m_i}{\|r - r_i\|}$$

where $\|r - r_i\|$ is the distance from field point *r* to mass point m_i.

In another instance, in the simplest electrostatic field created by a single point charge with charge *Q*, the electric potential or voltage at any location in the field can be calculated as follows, if the potential at an infinite distance is supposed to be zero,

$$\varphi(r) = \frac{Q}{4\pi\varepsilon_0 \|r\|}$$

where

  r is the radial coordinate of a point in the spherical coordinate system with its origin point *Q*
  ε_0 is the vacuum permittivity

If multiple point charges exist and the superposed electrical potential is linear, the electric potential at a point r for n point charges is equal to the sum of the potentials produced individually by every point charge,

$$\varphi(r) = \frac{1}{4\pi\varepsilon_0} \sum_{i=1}^{n} \frac{Q_i}{\|r - r_i\|}$$

where $\|r - r_i\|$ is the distance from field point r to point charges Q_i.

Another example is the central force field generated by a single nucleon, in which the potential value at any point in space can be calculated by the following formulas:

Square well potential: $\quad \varphi(r) = \begin{cases} V_0 & \|r\| \leq R \\ 0 & \|r\| > R \end{cases}$

Gauss potential: $\quad \varphi(r) = V_0 \cdot e^{-(\|r\|/R)^2}$

Exponential potential: $\quad \varphi(r) = V_0 \cdot e^{-(\|r\|/R)}$

where V_0 and R indicate the strength and range of the nuclear force, respectively.

From the above-mentioned physical fields, we can see that the potential at any point in space is directly proportional to the measurement of the strength of an object's interaction, such as the mass or charge, and it decreases as the distance to the source increases. In the case of gravitational and electrostatic fields, the potential is inversely proportional to the distance of the source and tends to be zero as the distance becomes infinity. The gravitational force, as the gradient force with potential function is, in essence, a long-range force, which decreases slowly with an increase in distance and which exerts an influence over an arbitrarily large distance. By contrast, in the case of a nuclear force field, the potential falls off much faster as the distance increases, and the nuclear force quickly decreases to zero, which indicates a short-range field. Moreover, in the above fields, the potential at any point in space is isotropic, which means the potential is isotropic, and the associated force field can be modeled as a spherical force field.

Inspired by this field theory in physics, we introduced the interaction between particles and their field model into the abstract data space to describe the relationship between data objects or sample points, and to reveal the general characteristics of the underlying data distribution.

Given a set of samples $D = \{x_1, x_2, \ldots, x_n\}$, each sample can be measured by p observation properties or variables. So that the jth observation property of the ith sample $x_i(i = 1, 2, \ldots, n)$ is $x_{ij}(j = 1, 2, \ldots, p)$, and the p observation properties of samples x_i can be written as a vector:

$$x_i = \left(x_{i1}, x_{i2}, \ldots, x_{ip} \right)', \quad i = 1, 2, \ldots, n$$

If we take each sample observation vector as a data point, then n samples constitute data points in space featured by p-dimensions. Provided each data object can be considered as a virtual object or mass point with a field around it, and the interaction of all data objects form a data field in the space, and any object located within the field is subject to the combined action of other objects, one data field may be determined in the entire space, namely, that formed by the data set $D = \{x_1, x_2, \ldots, x_n\}$ in p-dimensional featured space.

As for time-independent data sets, the associated data field can be regarded as a constant source field, which can be analytically modeled by vector intensity and scalar potential functions. Considering that it is inherently simpler to calculate scalar potential than vector intensity, we prefer to use scalar potential function to describe a data field.

4.1.3 Potential Field and Force Field of Data

According to field theory in physics, the potential of a constant field is an isotropic uniform function about position in the field, which is inversely proportional to the distance, and directly proportional to the magnitude of the particle's mass or charge. Thus, we can offer the following principles and set up some general rules for the potential function of a data field [2].

Given a data object x in p-dimensional space Ω, let $\varphi_x(y)$ be the potential at any point $y \in \Omega$ in the data field produced by x, then $\varphi_x(y)$ must meet all the following rules:

1. $\varphi_x(y)$ is a continuous, smooth, and finite function in space Ω.
2. $\varphi_x(y)$ is isotropic in nature.
3. $\varphi_x(y)$ monotonically decreases with the distance $\|x - y\|$.

When $\|x - y\| = 0$, $\varphi_x(y)$ reaches maximum, but is not infinity, and when $\|x - y\| \to \infty$, $\varphi_x(y) \to 0$.

In principle, all functional forms subject to the above rules can be used to define the potential function of a data field. The distance $\|x - y\|$ can be measured with norm distance L_q, that is

$$\|x - y\|_q = \left(\sum_{i=1}^{p} |x_i - y_i|^q \right)^{(1/q)}, \quad q > 0$$

When q is 1, 2, and ∞, respectively, the norm distance L_q is

$$\textit{Absolute value of the distance:} \quad \|x - y\|_1 = \sum_{i=1}^{p} |x_i - y_i|$$

$$Euclidean\ distance:\ \|x - y\|_2 = \sqrt{\sum_{i=1}^{p} (x_i - y_i)^2}$$

$$Chebyshev\ distance:\ \|x - y\|_\infty = \max_{1 \le i \le p} |x_i - y_i|$$

Usually, the Euclidean distance of $q = 2$ is used. Referring to the potential functions of the gravitational field and the nuclear force field, we propose two optional potential functions for a data field.

Potential function of a gravitational-like data field:

$$\varphi_x(y) = \frac{m}{1 + \left(\|x - y\|/\sigma\right)^k}$$

Potential function of a nuclear force-like data field:

$$\varphi_x(y) = m \times e^{-\left(\|x-y\|/\sigma\right)^k}$$

where $m \ge 0$ represents the strength of the field source, which can be regarded as the mass of an object or particle, and is usually used to describe the inherent property or importance of the data samples; $\sigma \in (0, +\infty)$ is used to control interactions between objects, called influence coefficient; $k \in N$ is the distance index.

Figure 4.2 presents the curves of the two potential functions in the case of m, $\sigma = 1$. In Figure 4.2a, with $k = 1$, the potential function of the gravitational-like field decreases slowly as the distance increases, representing a long-range field. In Figure 4.2b, with $k = 5$, both potential functions sharply decrease to zero, indicating they are short-range fields. Generally, the potential function representing a short-range field will better describe the interaction of data objects.

Considering the relationship between the functional forms and the distribution properties of associated potential fields, as shown in Figures 4.3 and 4.4, each equipotential line plot depicts a potential field created by the same 390 data points of unit mass, with different functional forms and different parameter settings. Let each data point have the same importance. From the figure, we see that different potential function forms lead to a different distribution of data field equipotential lines, but when the influence coefficient σ is valued at appropriate intervals, different potential function forms correspond to very similar potential field distributions.

Let us take the nuclear force-like functional form as an example for a further discussion of the influence range of a single data object of unit mass with respect to various k. As illustrated in Figure 4.5, in the case of $k = 2$, potential function

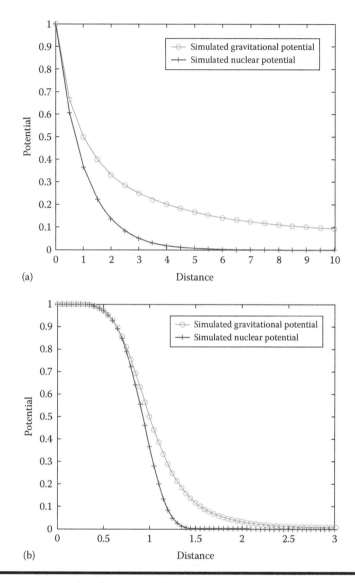

(a)

(b)

Figure 4.2 Two optional potential function forms (m, σ = 1): (a) k = 1 and (b) k = 5.

$\varphi_x(y) = m \times e^{-(\|x-y\|/\sigma)^2}$ corresponds to a Gaussian-like potential. According to 3σ law of Gaussian function, each data object has an influence vicinity of radius $\left(3/\sqrt{2}\right)\sigma$. The range of object interaction will gradually decrease as the order of distance term k increases. When $k \to \infty$, potential $\varphi_x(y)$ will approximate to a square well potential of width σ, that is, the range of object interaction tends to be σ. Consequently,

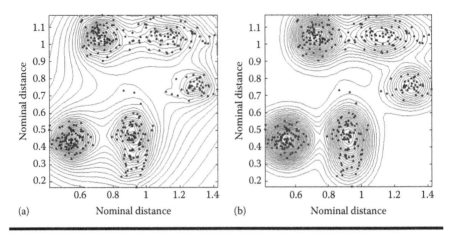

Figure 4.3 Influence of gravitational field potential function parameters on the spatial distribution of potential field data: (a) $k = 1$, $\sigma = 0.009$ and (b) $k = 5$, $\sigma = 0.095$.

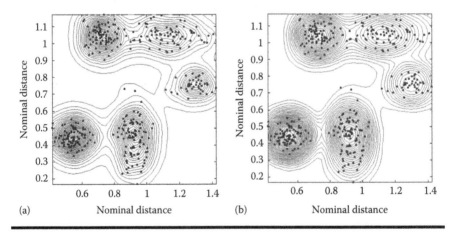

Figure 4.4 Influence of nuclear field potential function parameters on the spatial distribution of potential field data: (a) $k = 1$, $\sigma = 0.039$ and (b) $k = 5$, $\sigma = 0.129$.

a new measurement, called "influence radius," can be introduced to describe the range of interaction between objects in a data field:

$$R = \sigma \times \sqrt[k]{\frac{9}{2}}$$

Let us analyze the properties of the influence radius R of a single object data field, where R monotonically decreases in the order of the distance term k, if σ is fixed,

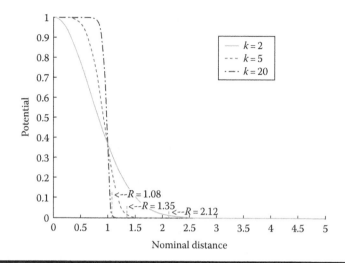

Figure 4.5 **Potential functions of a nuclear field with different *k* values and their influence radius (*m*, σ = 1).**

and $\forall k \in N$, $\sigma < R \leq (9/2)\sigma$. For example, the influence radii in Figure 4.4a and b are $R_a \approx 0.176$ and $R_b \approx 0.174$, respectively, indicating that data potential fields with a similar spatial distribution have approximately equal influence radii.

Thus, the spatial distribution of the data field is mainly determined by the interaction force between the objects or the influence radius, and there is no obvious correlation between the specific shape of the potential function and the value of distance index *k*. According to this discussion, since a Gaussian function has been widely applied for its good mathematical properties, we prefer to use it (i.e., nuclear force-like potential function with *k* = 2), to model the distribution of data fields involving human cognitive behaviors.

Given a data field produced by a set of data objects $D = \{x_1, x_2, \ldots, x_n\}$ in space $\Omega \subseteq \mathbf{R}^p$, the potential at any point $x \in \Omega$ can be calculated as

$$\varphi(x) = \varphi_D(x) = \sum_{i=1}^{n} \varphi_i(x) = \sum_{i=1}^{n} \left(m_i \times e^{-\left(\|x - x_i\|/\sigma\right)^2} \right)$$

where

$\|x - x_i\|$ is the distance between object x_i and point x

$m_i \geq 0$ is the mass of object $x_i (i = 1, 2, \ldots, n)$, which meets a normalization condition

$$\sum_{i=1}^{n} m_i = 1$$

Generally, each data object is supposed to have the same mass, that is, all objects have the same influence in the space, thus a simplified potential function formula can be given as

$$\varphi(x) = \frac{1}{n} \sum_{i=1}^{n} e^{-\left(\|x-x_i\|/\sigma\right)^2}$$

Just like a scalar field depicted with isolines or isosurfaces in physics, we use equipotential lines or surfaces to represent the spatial distribution of data potential fields. Given a potential ψ, an equipotential line or surface can be rendered where all points thereon satisfy the implicit equation $\varphi(x) = \psi$. Thus, by specifying a set of discrete potentials $\{\psi_1, \psi_2, ...\}$, a series of nested equipotential lines or surfaces can be drawn for a better understanding of the associated potential field as a whole.

As shown in Figure 4.6, the equipotential map for a potential field produced by a single data object consists of a family of nested, concentric, and data-centered object circles or spheres. The figure shows us that the equipotentials closer to the object have higher potentials and are further apart in distribution, which indicates strong field strength near the object. Another example shown in Figure 4.7 is a potential field created by 180 data points in a three-dimensional space. The same map cannot effectively show multiple potential surfaces, and corresponding potential fields can be represented by extracting the potential surface of a number of potential values. Obviously, equipotential surfaces corresponding to different potentials have quite different topologies. For instance, the equipotential at potential $\psi = 0.381$ is composed of many small topological components enclosing different data objects. As we

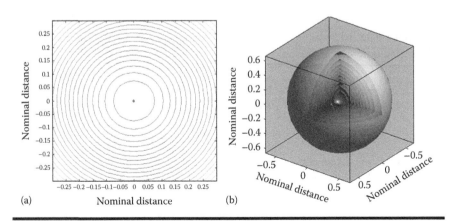

(a) Nominal distance   (b)

Figure 4.6 Distribution of equipotential line (surface) of single object data field ($\sigma = 1$): (a) isopotential map and (b) equipotential surface map.

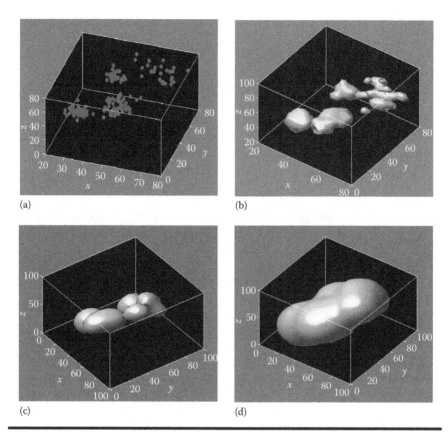

Figure 4.7 Equipotential surface distribution of potential field for three-dimensional data ($\sigma = 2.107$): (a) 180 data points in a three-dimensional space; (b) equipotential surface with potential value $\psi = 0.381$; (c) equipotential surface with potential value $\psi = 0.279$; and (d) equipotential surface with potential value $\psi = 0.107$.

go farther, small components will be nested into larger ones and the equipotentials start to enclose more and more data objects. When the potential equals to 0.279, the equipotential is composed of only five components comprising all the data objects. Eventually when the potential decreases to 0.107, all data objects "fall" inside an equipotential.

We will now explain the relationship between the potential function of data fields and nonparametric density estimations. According to the superposition principle of scalar potential, if data objects have equal mass, the denser regions of data will have higher potential, and vice versa. In this sense, potential function can be approximately considered as a density estimation of the underlying data distribution. Let $K(x)$ be a basis potential corresponding to $K(x) = e^{-\|x\|^k}$ and $K(x) = 1/(1 + \|x\|^k)$ for nuclear force-like and gravitation-like potentials, respectively, then the superposition of n basis potentials can be revised as,

$$\varphi(x) = \sum_{i=1}^{n} \left(m_i \cdot K\left(\frac{x - x_i}{\sigma} \right) \right)$$

According to the properties of probability density function, it can be proved that the difference between potential function and probability density function is only a normalization constant if $K(x)$ has finite integral in space $\Omega \subseteq \mathbf{R}^p$, that is, $\int_{\Omega} K(x) dx = M < +\infty$. In particular, if the data objects have equal mass, the overall nuclear force-like potential is quite similar to kernel density estimation in nonparametric density estimation. Given n independent and identically distributed d-dimensional samples $x_1, x_2, \ldots, x_n$, the kernel density estimation of overall density $p(x)$ can be written as

$$\hat{p}(x) = \frac{1}{n \times h^d} \sum_{i=1}^{n} \mu\left(\frac{x - x_i}{h} \right)$$

where
  $h > 0$ is its bandwidth
  $\mu(x)$ is the kernel function

In practical applications, the kernel function $\mu(x)$ usually chooses a symmetric density function with a unimodal peak at the origin, such as Gaussian kernel, uniform kernel, and so on. Essentially, kernel density estimation is the superposition of the basis kernel functions centered on each data point, and the estimated density at any point x in space is equal to the average contribution of each data point, which depends on the distance between the data point and point x. Obviously, if $K(x) = \mu(x)$ and each data object is equal in mass, the potential of a data field actually gives a physical explanation for the kernel density estimation, in which the requirements for a potential function are not as strict as that for the kernel density estimation. For example, basis function $K(x)$ is not required to be a density function, and each basis function must not have equal weight or mass.

Due to the fact that the vector intensity can always be constructed based on scalar potential by gradient operator, the vector intensity $F(x)$ at any point $x \in \Omega$, can be given as,

$$F(x) = \Delta\varphi(x) = \frac{2}{\sigma^2} \sum_{i=1}^{n} \left((x_i - x) \cdot m_i \cdot e^{-\left(\|x - x_i\| / \sigma \right)^2} \right)$$

Since the term $2/\sigma^2$ is a constant, the above formula can be simplified as,

$$F(x) = \sum_{i=1}^{n}\left((x_i - x) \cdot m_i \cdot e^{-\left(\|x-x_i\|/\sigma\right)^2} \right)$$

The field strength function of a data force field produced by a single data object is analyzed in Figure 4.8. It reaches its maximum value at the distance $r = 0.705$ and quickly decreases to zero as the distance exceeds $R \approx 2.121$, indicating a strong force on the sphere with radius = 0.705.

Figure 4.9 shows the distribution of force lines for a data force field produced by a single object ($m, \sigma = 1$) where the radial field lines are always on the spherical coordinate, directing toward the source, and perpendicular to the equipotentials, which indicates a gravitational force. The force lines are uniformly distributed and are spherically symmetric on any equipotential surface. The norm of force line reaches its maximum value at the positions with the radial distance = 0.705 and decreases as the radial distance further increases, which indicates a strong gravitational force on the sphere centered on the object and with the radius of 0.705.

Figure 4.10 shows the distribution of force lines for a data force field generated by 390 data points in two-dimensional space. The force lines always point toward the five local maximums of the associated potential field, which could be considered a data field produced by five "virtual point sources." Due to the attraction of the "virtual sources," all the data objects in the field will tend to converge toward each other and exhibit certain features of self-organization. This is the same for the data field produced by 280 data points in three-dimensional space shown in Figure 4.11.

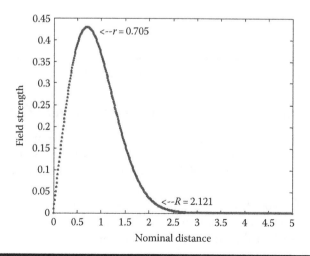

Figure 4.8 Field strength function of single object data field ($m, \sigma = 1$).

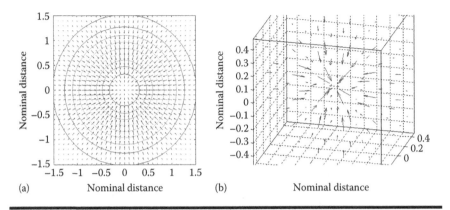

Figure 4.9 Field lines distribution of single object data field (m, $\sigma = 1$): (a) two-dimensional vector graphics and (b) three-dimensional vector graphics.

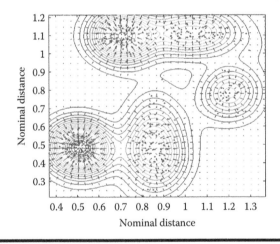

Figure 4.10 Field line distribution of two-dimensional data force field ($\sigma = 0.091$).

4.1.4 Selection of Influence Coefficient in Field Function

Given a data set $D = \{x_1, x_2, \ldots, x_n\}$ in space Ω, the distribution of the associated data field is primarily determined by the influence coefficient σ once the form of potential function is fixed. Examples of different potential fields produced by variances of σ are shown in Figure 4.12, where the fields are generated by five data objects in the form of a Gaussian potential. If σ is too small, the range of interaction is very short, and potential function $\varphi(x)$ will become the superposition of n peak functions centered on the data objects. The extreme condition is that no interaction exists between the objects, and the potential at the location of each data object nearly equals $1/n$.

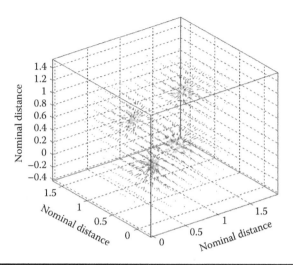

Figure 4.11   Field lines distribution of 3D data field (σ = 0.160).

As shown in Figure 4.12c, if σ is very large there is a strong interaction between objects, and φ(x) will become the superposition of *n* broad and slowly changing basis functions. The extreme condition is that the potential at the location of each object approximates to one. Since the difference between the probability density function and the potential function in the form of a Gaussian function is only a normalization constant, obviously, the potential in the above extreme cases cannot produce a meaningful estimation of the underlying distribution. Thus, the choice of σ should make the distribution of the potential field consistent with the underlying distribution of the original data, as much as possible.

Entropy is used to measure the amount of thermal energy, showing the disorderliness or randomness in a closed thermodynamic system. However, Shannon's entropy is a useful measurement of uncertainty in an information system. The higher the entropy is, the more uncertain the associated physical system. For the data field generated by the data objects $x_1, x_2, \ldots, x_n$, if the potentials at the positions of all objects are equal to one another, we are most uncertain about the underlying distribution. Shannon's entropy is highest in this case. Conversely, if the distribution of the potential field is highly skewed, both uncertainty and entropy will be low. Thus, we introduce potential entropy to measure the uncertainty in distribution of a potential field [2–5].

Let $\psi_1, \psi_2, \ldots, \psi_n$ be the potentials of objects $x_1, x_2, \ldots, x_n$, respectively. The potential entropy can be defined as

$$H = -\sum_{i=1}^{n} \frac{\psi_i}{Z} \ln\left(\frac{\psi_i}{Z}\right)$$

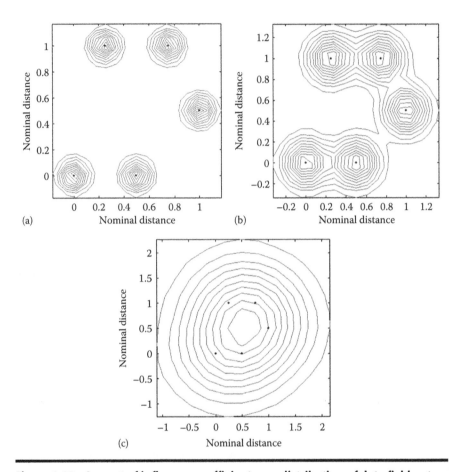

Figure 4.12 Impact of influence coefficient σ on distribution of data field potential: (a) σ = 0.1, (b) σ = 0.2, and (c) σ = 0.8.

where $Z = \sum_{i=1}^{n} \psi_i$ is a normalization factor. For any $\sigma \in [0, +\infty]$, the potential entropy H satisfies $0 \leq H \leq \log(n)$ and $H = \log(n)$, if and only if $\psi_1 = \psi_2 = \cdots = \psi_n$.

Figure 4.13a illustrates an example of the optimal selection of influence coefficient σ for a data field produced by 400 data points. Obviously, when $\sigma \to 0$, potential entropy H tends to $H_{max} = \log(400) \approx 5.992$; H decreases at first as σ increases from 0 to ∞; and at a certain $\sigma(\sigma_{opt} \approx 0.036)$, H achieves a global minimum $H_{min} \approx 5.815$; thereafter, H increases as σ further increases, and tends to be maximum again when $\sigma \to \infty$. Figure 4.13b demonstrates the data potential field with optimal σ, which obviously fits the underlying distribution very well.

In nature the optimal choice of σ is a minimization problem of a univariable, nonlinear function H, that is, $\min H(\sigma)$, which can be solved by standard optimization algorithms, such as a simple trial and error method, a stochastic searching

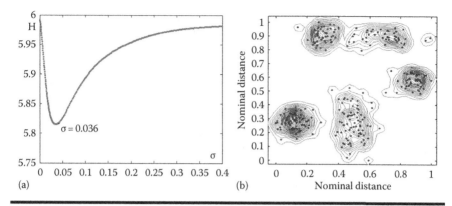

Figure 4.13 **Selection of influence coefficient σ: (a) relation curve for *H* and σ and (b) optimal the potential field distribution corresponding to σ.**

method, a simulated annealing method, and so on. In practical applications we can use random sampling techniques with sample size $n_{sample} \ll n$ to reduce the computational cost. Due to random sampling at the cost of distribution of the original data, the selected sample size n_{sample} should try to keep the inherent structure of the original data distribution. Generally speaking, in the case that *n* is very large, the sampling rate should not be less than 2.5% in order to keep the underlying distribution properties of the original data. Unless otherwise specified, the value of the influence coefficients mentioned hereinafter is calculated by an optimization algorithm based on potential entropy, and specific optimization methods will no longer be described.

4.2 Clustering Based on Data Field

4.2.1 Uncertainty in Classification and Clustering

Birds of a feather flock together and classification is one of humanity's most fundamental and significant social, production, and research activities. The basic ability of humans to understand the world comes through distinguishing different objects and classifying them according to their similarities and differences.

Classification induces the connotation of a concept according to its given extension, that is, to describe and distinguish models or classification functions and to use them to classify objects with an unknown class index. The induced models and functions are based on an analysis of the training data set with known class indices, so it is a supervised learning method.

Clustering is much more difficult than classification. Different clusters cannot be classified in advance, nor can they be given clear category tags, since there are no clear connotations nor denotations of the concepts. We cannot predetermine the number of categories, but need to identify the class mark of the data according to the actual distribution of the sample. By gathering the extension objects within the newly generated concepts, we then make the similarity between the classes as small as possible and that within the class as large as possible. There are more uncertainties in clustering, but it is a basic scientific question that is encountered in all disciplines.

For a given set of data samples, the number of classes and the center of each class depend not only on the distribution of actual data, but also on the background and purpose of the clustering applications. Cognitive psychology research shows that humans do not cluster too many classes at the same level, usually no more than seven. If there are more or less than seven classes, we can adjust the concept's granularity to re-cluster at a higher or lower level. Therefore, in this sense, there is no absolutely correct clustering division. In clustering, there is generally no training set or class labels, and each object contributes to the actual clustering result. Therefore, clustering is more complex than classification and contains more uncertainties.

According to the similarity measurement and the evaluation criteria of clustering, commonly used clustering methods include partitioning, hierarchical, density- and grid-based, and so on. We now propose a novel data field-based clustering method.

4.2.2 Dynamic Clustering Based on Data Field

The basic idea of dynamic clustering based on data force fields is introduced to describe interactions between objects and to cluster objects via a simulation of their motion in the data force field. Without effects from external forces, objects will be attracted by, and converge on, each other; when two objects meet, or are close enough, they will combine into a new object, whose mass and momentum are the sum of the first two; the object combination in the simulation can be regarded as a cluster combination in a hierarchical clustering process, that is, the sub-clusters are gradually combined into a new cluster at higher hierarchies with larger granularities; the algorithm is operated iteratively until all the object clusters go into one group.

In human cognition processes, more cluster numbers do not mean better clustering. For thousands, or more, collected data samples, each data sample does not need to be seen as an independent category with the same mass for each layer by layer combination. Hence, rather than displaying all the interactions and motions between all the objects, we first estimate the object mass, select a few representative kernel objects with non-zero mass to pre-cluster the data, then combine the representatives iteratively to implement a hierarchical clustering.

4.2.2.1 Selecting Representative Objects through Mass Estimation

Let $D = \{x_1, x_2, \ldots, x_n\}$ be a data set with n objects in the space $\Omega \subseteq \mathbf{R}^p$. The potential function of its data field is

$$\varphi(x) = \sum_{i=1}^{n} \left(m_i \times e^{-\left(\|x - x_i\|/\sigma\right)^2} \right)$$

If the mass, $m_1, m_2, \ldots, m_n$, is not be equal, they can be regarded as a group of functions in terms of spatial position $x_1, x_2, \ldots, x_n$. Once the overall distribution is obtained, the mass can be optimally estimated via minimizing the error criteria related to potential function $\varphi(x)$ and density function of the overall distribution. In detail, let $x_1, x_2, \ldots, x_n$ be n simple samples from a continuous d-dimensional set, and $p(x)$ be the overall density. When σ is fixed, we can minimize the squared integral of the error, such that

$$\min J = \min_{\{m_i\}} \int_{\Omega} \left(\frac{\varphi(x)}{\left(\sqrt{\pi}\sigma\right)^d} - p(x) \right)^2 dx$$

Expand this formula, and substitute $\varphi(x) = \sum_{i=1}^{n} \left(m_i \times e^{-\left(\|x - x_i\|/\sigma\right)^2} \right)$ in the formula, to obtain

$$\min J = \min_{\{m_i\}} \left(\frac{1}{2 \cdot \left(\sqrt{2}\right)^d} \sum_{i=1}^{n} \sum_{j=1}^{n} m_i \times m_j \times e^{-\left(\|x_i - x_j\|/\left(\sqrt{2}\sigma\right)\right)^2} \right.$$
$$\left. - \frac{1}{n} \sum_{i=1}^{n} \sum_{j=1}^{n} m_i \times e^{-\left(\|x_i - x_j\|/\sigma\right)^2} \right)$$

This is a typical constrained quadratic programming problem, satisfying the linear constraints $\sum_{i=1}^{n} m_i = 1$ and $\forall i, m_i \geq 0$. A group of optimal estimation of mass can be obtained through an optimal solution, denoted as $m_1^*, m_2^*, \ldots, m_n^*$ to satisfy the normalization condition. The function optimization aims to have a small number of objects in the dense region with larger mass, and most distant objects with smaller mass. In other words, the spatial distribution of the data field depends on the interaction between objects with larger mass, most other objects can be ignored due to their small contribution to field formation as a result of having too small a mass.

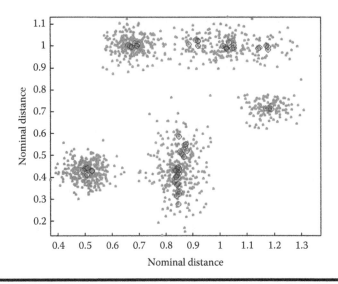

Figure 4.14 Simplified estimation of mass of 1200 objects in the data field (σ = 0.078).

For convenience, we will regard the objects with estimated non-zero mass as representative. Figure 4.14 shows the estimated results for the initial 1200 data sample objects in a two-dimensional space. Given the mass threshold, according to the above estimation calculation method, each object can be assigned a value of mass, and 71 representatives of larger mass can be selected, marked with a circle. The number of representative objects and the selection process are determined by the field distribution of the original data; although other objects are discarded, their contribution to the clustering has been reflected in the mass value of the representative objects. Further comparison of the data field isopotential maps generated by the representative object set and the original datasets, as shown in Figure 4.15, shows that both are characterized by a very similar potential field distribution. That is, the data field generated by the original data set is well approximated with the data field generated by a small number of representative objects; or the spatial distribution of the data field depends primarily on the interaction between representatives of a larger mass, while the effect on the data field of most other objects with a small mass can be ignored.

4.2.2.2 Initial Clustering of Data Samples

Let the set of representative objects $D_{rep} \subseteq D$. We regard each representative object as the center of one class in the initial clustering. All objects attracted by the same representatives are considered one data cluster. Thus, we can obtain the initial cluster based on representatives, which can be mathematically described as follows.

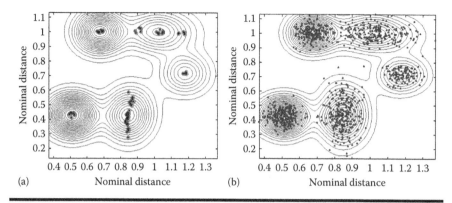

Figure 4.15 **Comparison of data fields generated by representative objects and by the original data set ($\sigma = 0.078$): (a) 71 representatives object and (b) the original data set.**

Let the representative and its mass be $x^* \in D_{rep}$ and m^*, respectively. If a subset $C \subseteq D$ exists, $\forall x \in C$ is in the same point sequence $x_0 = x, x_1, \ldots, x_k \in \Omega$, satisfying

$$\left\| x_k - x^* \right\| < 0.705 \sigma m^*$$

And if x_i is on the gradient direction of $x_{i-1} (0 < i < k)$, then C is the cluster with x^* as center.

To be specific, other objects are classified into corresponding representative objects via a climbing method instructed by the field strength direction. As shown in Figure 4.14, the 1200 data samples are allocated to the 71 class centers to form 71 initial classes. The hierarchical combination of the initial classes can be achieved by the interaction between and the movement of representative objects.

4.2.2.3 Dynamic Clustering of Representative Objects

Interactions between representative objects and their further consolidation are simulated by the movement of each representative object in the time period $[t, t + \Delta t]$. Let $m_i^*(t)$ denote the mass of the representative $x_i^* \in D_{core}$ at the time t, and $x_i^*(t)$ denote the position vector, $i = 1, 2, \ldots, |D_{core}|$. In the case of there being no external forces, the field force and the instantaneous acceleration of the object in the data field are

$$F^{(t)}\left(x_i^*\right) = m_i^*(t) \times \sum_{x_j^* \in D_{rep}} \left(m_j^*(t) \times \left(x_j^*(t) - x_i^*(t)\right) \times e^{-\left(\frac{\left\| x_j^*(t) - x_i^*(t) \right\|}{\sigma}\right)^2} \right)$$

$$a^{(t)}\left(x_i^*\right) = \frac{F^{(t)}\left(x_i^*\right)}{m_i^*(t)}$$

If Δt is small enough, each representative object can move at an approximately uniform variable in the interval $[t, t + \Delta t]$. Let the initial velocity vector of each iteration be 0, then the position of the object at $t + \Delta t$ is

$$x_i^*\left(t + \Delta t\right) = x_i^*\left(t\right) + \frac{1}{2} a^{(t)}\left(x_i^*\right) \times \Delta t^2$$

When two representative objects meet, or their space

$$\left\| x_i^*(t) - x_j^*(t) \right\| \leq 0.705\sigma \times \left(m_i^*(t) + m_j^*(t)\right)$$

then they merge into a new representative object, x_{new}^*. According to the law of conservation of momentum, the mass and position of the new object are calculated as

$$m_{new}^*(t) = m_i^*(t) + m_j^*(t)$$

$$x_{new}^*(t) = \frac{m_i^*(t) \times x_i^*(t) + m_j^*(t) \times x_j^*(t)}{m_i^*(t) + m_j^*(t)}$$

Continue with the algorithm recursively until all representative objects are aggregated into one object or satisfy the termination condition.

Prior to each iteration, first calculate the representative's current minimum distance min_*dist* and maximum acceleration max_*a*, then let

$$\Delta t = \frac{1}{f}\left(\sqrt{\frac{2\min\_dist}{\max\_a}}\right)$$

where f represents temporal resolution.

The algorithm is described as follows:

Dynamic Clustering Based on Data Field Algorithm
Input: Dataset D.
Output: Hierarchical division of the data $\{\Pi_0, \Pi_1, \ldots, \Pi_k\}$.

Steps:

1. Generate a random subset $D_{sample} \subset D$, optimize the value of influence coefficient σ.
2. Solve the constrained quadratic programming problem to estimate the mass of the data object and take the object as a representative.
3. Take the representative objects as cluster centers, then form the initial data partition Π_0.
4. Iteratively simulate the interaction between representative objects, and further combine objects to form the hierarchical structure of the initial cluster and to get the clustering results $\{\Pi_0, \Pi_1, ..., \Pi_k\}$ at different levels.

Analysis of time complexity of the algorithm is as follows:

Step (2) simplifies the mass estimation of objects by solving the constrained quadratic programming problems with a sequential minimal optimization method, where the time complexity is $O\left(n_{sample}^2\right)$, $n_{sample} \ll n$.

In Step (3) all non-zero mass objects are allocated to the corresponding representative objects to realize an initial partition. Let the number of representative objects be n_{rep}, and the average time complexity be $O(n_{rep} \times (n - n_{rep}))$.

In Step (4) the representative objects are combined through attraction to each other, the average time complexity is $O\left(n_{rep}^2\right)$; the total time complexity of the algorithm is $O\left(n_{sample}^2 + \left(n - n_{rep}\right) \times n_{rep} + n_{rep}^2\right) \approx O(n)$.

To ensure the effectiveness of the clustering results, the selection of a sample size of n_{sample} should retain the distributive features of the original data. When n is very large, the recommended minimum sampling rate is usually not less than 5%.

[Experiment] The test data set in the literature [6] is used to test the effectiveness of the clustering algorithm based on the data field. This data set contains five clusters of different shapes, sizes, and densities, with some white noise data, totaling 100,000 data points, as shown in Figure 4.16. The compared algorithms are the popular K-means algorithm [7] and the improved hierarchical clustering algorithm BIRCH [8], CURE [6], and so on; all the programs are implemented in MATLAB®.

During the test, considering the high time complexity $O(n^2 \times \log n)$ of the CURE algorithm, we selected random 8000 data points to make five clusters, and analyzed the clusters with different algorithms, such as K-means, BIRCH and CURE algorithms, and a clustering algorithm based on the data fields. Figure 4.17 demonstrates the clustering results.

A spherical basis exists in BIRCH and K-means. The CURE clustering algorithm uses multiple representative points with good separation properties to present clustering; while, according to the specified shrinking factor α, the CURE algorithm absorbs other points to the center, so as to effectively suppress the influence of noise or outlier points in the clustering boundary. CURE can discover the correct clustering structure in the distribution if appropriate parameters are selected, but it cannot

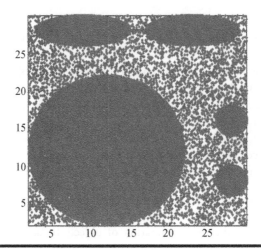

Figure 4.16 Test data composed of 100,000 data points.

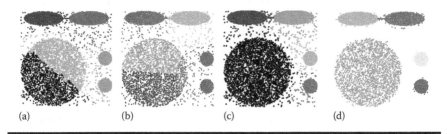

(a) (b) (c) (d)

Figure 4.17 Comparison of the clustering results (*k* = 5, different colors to represent different classes): (a) BIRCH, (b) K-means, (c) CURE, and (d) data-field clustering.

effectively deal with the noise data. In addition, the effectiveness of the clustering algorithm is heavily dependent on the number of representative points and the value of shrinking factor α, as shown in Figure 4.18. If α is too large, there is spherical bias in the clustering result; whereas, if α is too small, the clustering result is susceptible to noise or outlier data. In practical applications it is very difficult to make users select the appropriate algorithm parameter values.

Comparatively, a dynamic clustering algorithm based on data fields not only features a satisfactory clustering quality and effective processing of noise data, but results are independent of any elaborate selection of parameters by the user. Figure 4.19 shows the representative objects with simplified mass estimation and the corresponding clustering result. It is clear that if the representative objects are taken as clusters of representative points, the clustering algorithm based on data fields actually uses multiple representative points to describe the clustering distribution characteristics. But, unlike the CURE algorithm, the representative points

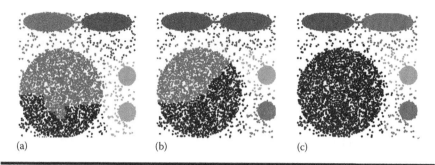

Figure 4.18 CURE algorithm relies heavily on the number of representative points and the shrinking factor values: (a) the representative point number of clustering is 10, $\alpha = 0.8$; (b) the representative point number of clustering is 5, $\alpha = 0.3$; and (c) the representative point number of clustering is 10, $\alpha = 0.8$.

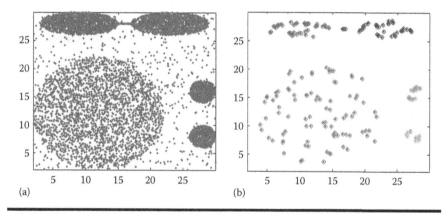

Figure 4.19 The original data set and 271 representative objects as a result of simplifying the estimation: (a) the experimental data set (8000 points) and (b) 271 representative objects as a result of simplifying the estimation ($\sigma = 0.6073$).

are selected by simplifying the estimation of the object mass, which can be better adapted to the clustering property of the underlying data distribution.

We further investigate the scalability of dynamic clustering by utilizing the test data set shown in Figure 4.16. Test results in Figure 4.20 show that the algorithm has good scalability, and the execution time of the algorithm and the size of the data set approximately present a linear relationship.

4.2.3 Expression Clustering of Human Face Images Based on Data Field

With the rapid development of the Internet, various forms of unstructured data keep emerging, such as graphics, images, audio data, and geospatial data. Therefore,

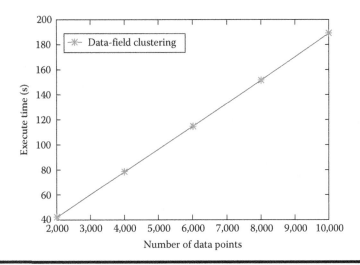

Figure 4.20 **Scalability of the data-field clustering algorithm with respect to size of the data set.**

one of the main challenges faced by data mining is the efficient handling of big unstructured data.

Unstructured face image data have become ubiquitous in daily life. Its recognition and understanding not only has a theoretical value, but could also enjoy broad application prospects. Human face recognition is essentially matching the identification of a two-dimensional image to a three-dimensional object. The main difficulties are [9]:

1. Face images often show a tilted head, a grimacing face, looking down or up, and other phenomena; the elastic deformation created by facial expressions, weight, and so on, have many uncertainties.
2. Face models, such as beard, hairstyle, glasses, make-up, and so on, make the recognition more difficult.
3. Image acquisition processes have many uncertainties, such as light intensity, number of light sources, light source direction, camera angles, and so on.

Face recognition is mainly based on those facial features that show big differences between individuals, but that remain relatively stable for the same individual. Each individual has unique facial features, and no two individuals are exactly the same, and even the same person has a different face at different times. Due to the limitation of objective conditions and the complexity of face changes, it is very difficult to represent and extract facial features. So far, there is no complete feature set suitable for face recognition, nor is there a universally applicable feature extraction method.

Face recognition has a rich content, which can identify age, race, face, personality, and even close relations, in addition to the identity of a specific person.

Expression recognition is a very difficult research topic within face recognition. The face is composed of 43 muscles whose subtle changes can present countless expressions. People believe there are seven basic emotions, expressed as anger, happiness, surprise, sadness, disgust, fear, and neutral. Most of the commonly used facial expression recognition methods identify faces by analyzing the muscle movements involved in each expression. We see facial expression recognition as a cognitive process of finding common or different knowledge in different face image data, and present a method to recognize face image expression based on data fields. The research starts with a frontal face gray-scale image, in which each pixel in the gray-scale image is regarded as a data object, and the gray-scale of the pixel as the object mass. Then we use the feature extraction method based on the data field to extract a small number of important pixels to form the featured face image. Next we use the Karhunen–Loeve Transform to extract the primary eigenvector of the face image set, and project the featured face image on to the common feature space spanned by the primary eigenvector to form the second-order data field. This ends in the recognition of a facial expression cluster based on the data field.

The Japanese female facial expression database (JAFFE) [10] was used for tested data, which consists of 213 frontal facial gray-scale images of 10 people. Each original image has a size of 256 × 256 pixels, and 256 levels of grayness. Figure 4.21 shows 21 frontal face gray-scale images of the same person with different expressions, including anger, happiness, surprise, sadness, disgust, fear, and a neutral expression.

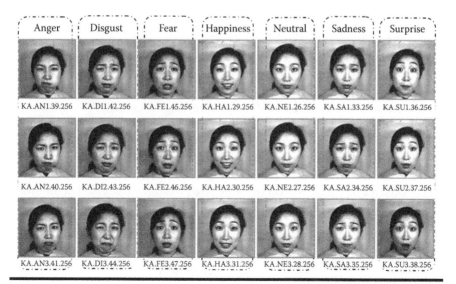

Figure 4.21 Twenty-one tested face images, 256 × 256 pixels, taken from JAFFE database.

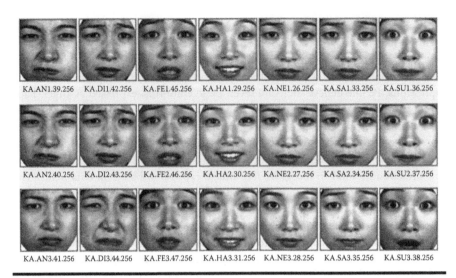

Figure 4.22 Post-preprocessing, 21 standard face images of 64 × 64 pixels.

For the test the original images were normalized via preprocessing. Rotation, cutting, and scaling were applied to each image to eliminate any disturbance from hairs, background, and illumination. After preprocessing, we obtained standard facial images of 64 × 64 pixels, in which the location of the two eyes was fixed, as shown in Figure 4.22. The feature extraction method based on data field was then used to extract important feature points from the image.

4.2.3.1 Feature Extraction Based on Face Image Data Field

In a given $m \times n$ dimensional face gray-scale image $A_{m \times n}$, if each pixel point is regarded as a data object in a two-dimensional space and the gray value $A(i, j)$ of the pixel is considered as the mass of the data object, the interaction among all pixels in the two-dimensional image space can identify a face image data field. For convenience, the gray-scale image $A_{m \times n}$ is represented as a vector for the formation of a stack row, that is, in a given gray-scale face image $A_{m \times n}$ with size of $m \times n$ pixels, if each pixel is considered as a data object in a two-dimensional image space, and the associated gray value $\rho_{ij} = A(i, j)(i = 1, 2, \ldots, m, j = 1, 2, \ldots, n)$ is considered as the mass of associated data objects, the interactions between all the pixels will form a data field in the image space. For convenience, $A_{m \times n}$ is represented as a vector of line stacking, that is,

$$B = (q_j) = \begin{bmatrix} \rho_0 \\ \rho_1 \\ \vdots \\ \rho_{m-1} \end{bmatrix}, \quad \text{which elements } \rho_i = \begin{bmatrix} \rho_{i0} \\ \rho_{i1} \\ \vdots \\ \rho_{i,n-1} \end{bmatrix}$$

where

B is $m \times n$-dimensional column vector, that is, $j = 1, \ldots, (m \times n)$

ρ_i is column vector of ith line in image $A_{m \times n}$, and ρ_{ij} is assumed to have been normalized into the interval $[0, 1]$

According to the potential function formula of the data field, the potential at any point x can be calculated as

$$\varphi(x) = \sum_{i=1}^{m \times n} \left(q_i \times e^{-\left(\|x - x_i\| / \sigma \right)^2} \right)$$

where σ is the influence coefficient used to control the influence range of each pixel. An equipotential line chart is used to describe the spatial distribution of the face image data field. Figure 4.23a shows a 64 × 64 pixels standardized facial image, whose equipotential lines distribution of data field and three-dimensional view are shown in Figure 4.23b and c. The high-potential area is located in the cheek, forehead, and

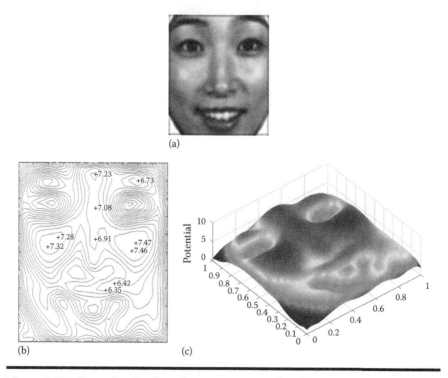

Figure 4.23 Data distribution of KA.HA2.30.256 face image ($\sigma = 0.05$): (a) standardized face image, (b) equipotential lines distribution of the data field, and (c) three-dimensional view of the two-dimensional potential field.

nose areas with large gray value. In other words, the potential field distribution of face image data can highlight cheeks, foreheads, noses, and other facial areas.

Considering that the eyes, mouth, eyebrows, nasal root, and other facial organs are important facial features in recognizing a frontal face image, such organs usually correspond to the low gray areas of the face image. To highlight the low gray area, the gray-scale of each pixel in the face image is first transformed nonlinearly $f(q_i) = (1 - q_i)^2$. According to the potential function formula of the data field, the potential value of any pixel x is

$$\varphi(x) = \sum_{i=1}^{m \times n} \left(f(q_i) \times e^{-\left(\|x - x_i\|/\sigma\right)^2} \right)$$

where σ is the influence coefficient that reflects the influence range of each pixel.

Figure 4.24a shows a 64 × 64 pixels standardized facial image, with post-nonlinear transformed equipotential lines distribution of data field, and three-dimensional

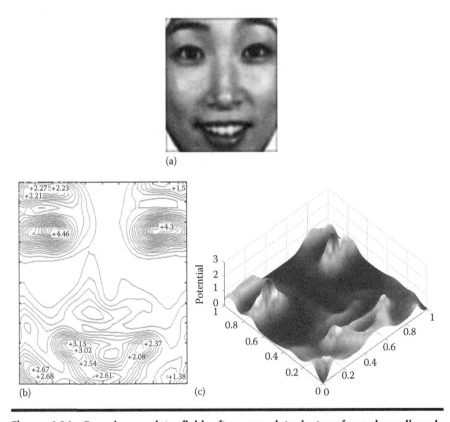

Figure 4.24 Face image data field after gray data is transformed nonlinearly (σ = 0.05): (a) standard face image, (b) distribution of equipotential lines of the data field, and (c) three-dimensional view of the two-dimensional potential field.

view shown in Figure 4.24b and c. Obviously, according to the local maxima value points of the potential function, we can accurately locate the eyes, eyebrows, mouth, and other important facial features.

We suppose that the facial image data field is mainly determined by the interaction between significant feature points. A weight variable $w_{ij} \in [0, 1]$ can be introduced to measure the contribution of each pixel to the formation of facial data field, where w_{ij} is assumed to satisfy the constraint $\sum_{i=1}^{m} \sum_{j=1}^{n} w_{ij} = 1$. According to the idea of optimal estimation in Section 4.2.2, we can get the following optimization objective function:

$$
\min_{\{w_i\}} \left(\frac{1}{2 \cdot \left(\sqrt{2}\right)^d} \sum_{i=1}^{m \times n} \sum_{j=1}^{m \times n} w_i \times w_j \times f\left(q_i\right) \times f\left(q_j\right) \times e^{-\left(\|x - x_{ij}\|/\left(\sqrt{2}\sigma\right)\right)^2} \right.
$$

$$
\left. - \frac{1}{m \times n} \sum_{i=1}^{m \times n} \sum_{j=1}^{m \times n} w_i \times f\left(q_i\right) \times e^{-\left(\|x - x_{ij}\|/\sigma\right)^2} \right)
$$

To optimally solve the objective function, the weight of each pixel in the face image needs to be obtained, and the pixels of non-zero weight are regarded as the feature points. Figure 4.25 gives 48 important feature points, using the feature extraction method based on the face image data field. The face image composed of these features not only describes the local geometric features represented by eyes, mouth, and other facial organs in happiness, surprise, anger, and other expressions, but also shows a good robustness to illumination changes.

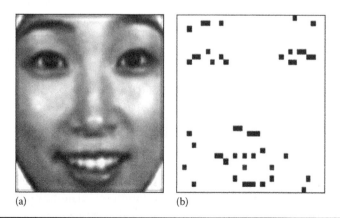

(a) (b)

Figure 4.25 Feature extraction based on face image data field ($\sigma = 0.05$): (a) standard face image and (b) 48 feature points extracted from standard face.

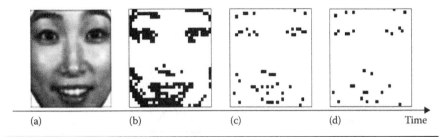

| (a) | (b) | (c) | (d) | Time |

Figure 4.26 **Oblivion—formalized description of memory loss: (a) a standard face image, (b) 250 feature points (σ = 0.02), (c) 48 feature points (σ = 0.05), and (d) 28 feature points (σ = 0.09).**

The influence coefficient σ herein is an important parameter variable. Figure 4.26 shows the effect of different σ values on feature point extraction. Different values lead to different results. When σ = 0.02, 250 feature points can be extracted, which reflects details in primitive face images; when σ = 0.09, 28 extracted feature points can generally describe the overall distribution of eyes, mouth, eyebrows, nose root, and other facial organs. In a sense, different feature point images corresponding to different σ values are like observing human faces from different distances, or a lens that can be zoomed in or out. A change in the observation distance can form face images of different sizes. The smaller σ is, the more feature points there will be, and the more detail for describing primitive face images; the bigger σ is, the fewer the feature points, and the more macro and broader the description of the face images. In practical applications, different σ values can be selected according to different requirements. If we want to extract features from all the standard face images in Figure 4.22 using the data field method, let σ = 0.05, then we get the face images based on important feature points, as shown in Figure 4.27.

The data field method that obtains a different number of feature points by choosing different σ values can also be used to formalize the description of memory loss. In the course of human cognition, oblivion is a performance of intelligence. Oblivion is the decline of memory, which is the process of the number of feature points of a human face reduced as time increases, as shown in Figure 4.26.

4.2.3.2 Recognition of Facial Expression Cluster Based on K–L Transformation and Second-Order Data Field

We can use the K–L transformation to further reduce the dimension of those feature face images. We can select the eigenvectors corresponding to the first p (p far less than total face image samples), using the largest feature values to construct a transformation matrix to form a common "feature face" space. Then we project each feature face image into the common feature face space to form projection data points. We can use the clustering method based on a data force field to cluster the data points in the feature face space to realize recognition of the expression cluster. Figure 4.28 shows the first two feature face images obtained by K–L transform, the clustering result is

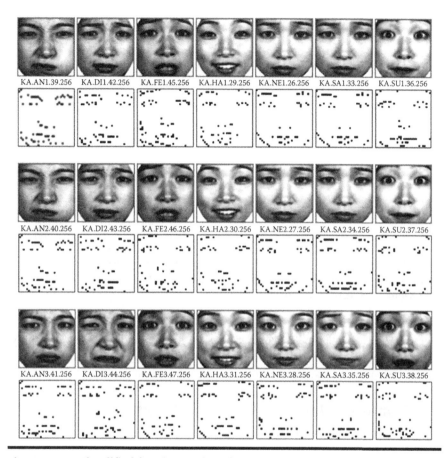

Figure 4.27 Simplified face images based on important feature points ($\sigma = 0.05$).

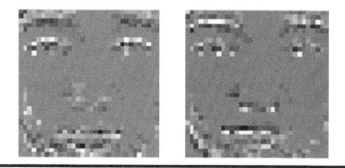

Figure 4.28 The first two "feature faces."

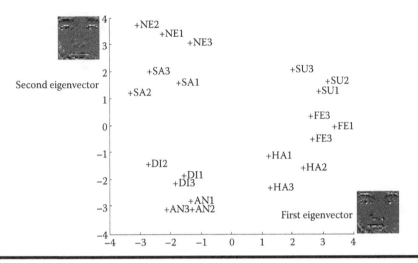

Figure 4.29 Projection of the test face image clustering results in the two-dimensional "feature face" space.

shown in Figure 4.29. Among them, AN, DI, FE, HA, NE, SA, and SU represent anger, disgust, fear, happiness, neutral expression, sadness, and surprise, respectively. It can be seen that the projection distribution of the feature face images, with different expressions in the feature face space, can be well separated.

4.3 Complex Network Research Based on Topological Potential

It is the basic idea of system science that structure determines function. If the various elements of the system are abstracted as nodes, and the relationship between the elements is regarded as connection, then the system constitutes a network with complex connections. There are a large number of complex networks in the real world, such as the Internet, power grids, transportation networks, personal relationships, cooperation networks, as well as neural networks, metabolic networks, and protein interaction networks in biological systems, and so on. Research shows that these seemingly unrelated real networks in different forms often share identical topological properties and are subject to certain basic rules of evolution. In 1998, Watts and Strogatz published a paper in *Nature* that described the small-world effect of real complex networks, namely that networks have a smaller average distance [11]. In 1999, Barabási and Albert published a paper in *Science*, pointing out that the degree distribution of many real networks follow the power-law distribution, called a scale-free network [12]. They attribute scale-free networks generated through self-organization by the real networks to two main factors, namely, the addition of nodes and the preferential attachment of connection between nodes. Many natural and social phenomena meet these two preconditions. At present, the term complex network has become a very common abstract description for various complex systems in technology, biology, and society. There has been extensive related research in the fields of physics, mathematics, biology, economics, sociology, and information science, thus boosting an important interdisciplinary research frontier.

At present, with the popularity of Internet and the development of network science, people are paying more and more attention to the evolutionary behavior of complex networks formed by communication and interaction between individuals capable of subject behavior. To understand the evolutionary behavior and organizational principles of these complex networks, we need to construct a mapping model from individual behavior to overall network structure. By learning from topology in mathematics and field theory in physics, we propose a topological potential method, that is, the construction of a virtual potential field in the network topological space. Because topological space is non-directional, it is impossible to draw a visualized potential field and equipotential line chart. However, we can use potential value to reflect the subjectivity of nodes, and the interaction between nodes, to study local influence and preferential attachment between individuals, and to reveal the structural evolution mechanism of complex networks.

4.3.1 From Data Field to Topological Potential

Inspired by the idea of data fields, we deem network G as an abstract system that contains n nodes and their interactions. A field exists around each node, and any node in the field will be co-affected by other nodes. Unlike the definition of data field, the effect of nodes in the network can only be transmitted through the side. The interaction between nodes is closely related to the node attribute and the network distance between nodes, in this way a potential field is determined in the whole network.

According to the modularity and clustering property of a real network, we believe that the interaction between nodes is local and the influence of each node will decay rapidly as distance increases. Therefore, we prefer to use a Gaussian potential function, which represents short-range fields and has good mathematical properties, to describe the interaction between nodes.

Given a network $G = (V, E)$, where $V = \{v_1, \ldots, v_n\}$ is a non-empty finite set of nodes, $E \subseteq V \times V$ is a collection of nodal coupling or side; according to the definition of potential function of data fields, the topological potential of any node $v_i \in V$ can be expressed as

$$\varphi(v_i) = \sum_{j=1}^{n} \left(m_j \times e^{-\left(d_{ij}/\sigma\right)^2} \right)$$

where

d_{ij} represents network distance or the number of steps between v_i and v_j, which is measured by the shortest path length herein; influence coefficient σ is used to control the impact range of each node

$m_j \geq 0$ represents the mass of node v_j ($j = 1, \ldots, n$), which can be used to describe the inherent properties of each node

In real networks, the inherent properties of a node have an abundant physical meaning, such as the city scale in an urban traffic network, the social background and the activity ability of an individual in the social network, the storage capacity of nodes in a communication network, and so on.

First, the difference in the inherent properties of nodes is neglected; let the mass of each node be equal and satisfy the normalized condition, so the simplified topological potential formula is

$$\varphi(v_i) = \frac{1}{n} \sum_{j=1}^{n} e^{-(d_{ij}/\sigma)^2}$$

According to the mathematical property of Gaussian function, for a given value σ, the impact range of each node is approximately equal to the local area where network distance is less than $\lfloor 3\sigma/\sqrt{2} \rfloor$. When the jump distance is greater than $\lfloor 3\sigma/\sqrt{2} \rfloor$, the unit potential function quickly decays to zero, indicating that it is a short-range field.

4.3.2 Important Network Nodes Detected with Topological Potential

In complex networks, the importance of nodes often varies. Usually a small number of key nodes with a large degree determine the behavior of the entire network, while the vast majority of the nodes have a relatively small degree, this constitutes a scale-free network. For this kind of non-uniform network, the degree value is obviously the key factor in evaluating the importance of nodes, but calculating the degree of a node only involves the connection of the node itself, which cannot effectively reflect difference or the importance of the position of a node in the overall topology. In fact, the importance of the node not only depends on the connection of node itself, but is also related to the importance of neighboring nodes. For example, a node may become very important because of its connection to another important node, or as the routing node of another local network. In addition, the inherent properties of a node also affect its importance in the network. Therefore, how to comprehensively consider the topological differences and the inherent properties of nodes so as to measure their importance in the network has become a primary problem in the study of complex networks, which is also a fundamental question in the analysis of social networks and information searches.

The PageRank algorithm for Internet searches is a very effective solution. While emphasizing the importance of node degree, it also considers the overall influence of the node's inherent properties, so the value for importance level is introduced. As a result, each node in the network influences other nodes through their degree and value for importance level. In turn, other nodes exercise the same influence on the node. But the PageRank algorithm does not carefully analyze the influence of different damping factor *d* on the importance ranking of nodes, but only gives an empirical value of 0.85.

On the basis of field data, we propose a method for evaluating node importance based on topological potential [13,14]. This method describes the interaction between nodes with potential, defines the topological potential of nodes, and depicts their different topological positions and importance as reflected by their own properties. The algorithm for ranking node importance based on topological potential involves three basic steps:

1. Optimally selecting influence coefficient σ
2. Calculating the topological potential of each node according to the optimal value of σ
3. Ranking according to the potential value to determine the importance of nodes

Similar to the optimization method of the influence coefficient σ in the data field, topological potential entropy *H* is introduced to optimize the value of σ

$$H(\sigma) = -\sum_{i=1}^{n} \frac{\varphi(v_i)}{Z} \ln\left(\frac{\varphi(v_i)}{Z}\right)$$

where $Z = \sum_{i=1}^{n} \varphi(v_i)$ is a normalization factor.

The optimization problem becomes the minimization problem of single-variable function $H(\sigma)$, that is, $\min H(\sigma)$.

Considering the long time needed for an iterative computation of node topological potential, an optimization interval of σ can be estimated first, and then its optimized value can be searched for precisely. The algorithm is described as follows:

Node Ranking Algorithm Based on Topological Potential
Input: Network $G = (V, E)$, where $V = \{v_1, \ldots, v_n\} |E| = m$;
Output: Output each node and its topological potential in descending topological potential value;
Steps:

1. Initialize potential entropy $H = \min\_H = -\sum_{i=1}^{n} \left(\deg(v_i)/(2m)\right) \ln\left(\deg(v_i)/(2m)\right)$, where $\deg(v_i)$ is the degree of node v_i.
2. Initialize 1-hop neighbor node set *lhop_neighbors*(v_i, l), $i = 1, \ldots, n$, that is, direct neighbor node set whose network distance is 1.
3. Let $l = 1$.

4. While $H \le \min\_H$ do
   {Let $l = l + 1; \sigma = \left(\sqrt{2}/3\right) l$

   According to the $l - 1$-hop neighbors of each node, calculate l-hop neighbor set $lhop\_neighbors(v_i, l)$, $i = 1, \ldots, n$;
   Calculate the topological potential of each node

   $$\varphi(v_i) = \sum\nolimits_{j=1}^{i} \left| lhop\_neighbors(v_i, j) \right| \times e^{-(j/\sigma)^2}, i = 1, \ldots, n;$$

   Calculate potential entropy $H = -\sum\nolimits_{i=1}^{n} \left(\varphi(v_i)/Z \right) \ln\left(\varphi(v_i)/Z \right)$.
   }

5. Let σ be in the search interval of $\left(\left(\sqrt{2}/3\right)(l-2), \left(\sqrt{2}/3\right) l \right)$, and search the minimum potential entropy that meets the accuracy requirement.

6. According to the topological potential distribution corresponding to the minimal potential entropy, output node in descending potential value.

Here is an analysis of the time complexity of the algorithm. The time complexity of steps (1) and (2) is $O(m)$; the time complexity of Step (4) is $O(m + n^{3/\gamma})$ in the best case, where $2 < \gamma < 3$ is a constant, and $O(n^2)$ in the worst case; the time complexity of Step (5) depends on time overhead for iterative calculation of potential entropy of all nodes. The calculation of potential entropy of each node involves only the number of neighbors that a node has in a given $l - 1$-hop, the time complexity is $O(ns)$, and s is the number of iterations. Therefore, the total time complexity of the algorithm is between $O(m + n^{(3/\gamma)})$ and $O(n^2)$, that is, the time complexity of the algorithm in the worst case is n^2, which is not high.

We shall now take the famous Zachary social network [15] as a typical case to check the validity of the topological potential algorithm and compare it with the algorithms for degree ranking, betweenness ranking, closeness ranking, and PageRank.

The Zachary social network is a small test network commonly used in the field of complex network and social network analysis. In the 1970s Wayne Zachary spent 3 years observing relations between members of a karate club in an American university and constructed the social relations network as shown in Figure 4.30. The network consists of 34 nodes and 78 edges, where nodes represent club members and lines between nodes represent occasions two members often appear together in other than club activities, which means that they can still be friends outside club activities. During the investigation, because of a dispute between club manager A. John node 34, and coach Mr. Hi node 1, two small groups formed with them as the core; nodes in square and circle represent different group members in Figure 4.30. As a real small social network, this network is often used to test the validity of node importance evaluation and community discovery.

We adopted the degree ranking, betweenness ranking, and closeness ranking methods, and the PageRank and topological potential algorithms, to compare the

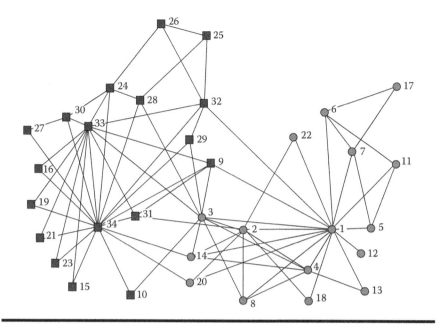

Figure 4.30 Zachary's karate club network (*n* = 34).

node importance for the Zachary network, as shown in Figure 4.31. In Figure 4.31b through f, the top 10 nodes are marked large circle; the larger the circle, the larger the node value, and the more important the node in the network topology. Specific values of a node in different metrics are presented in Table 4.1.

We made a further analysis of these ranking results. The 10 most important nodes are same with degree, PageRank and topological potential algorithms, but are different with betweenness and closeness ranking methods. The betweenness ranking method considers that node v_6 is more important than v_4, but the degree of node v_4 is larger than node v_6 in the network, closer to the center position of the network, and also connected with more important nodes, such as nodes v_1, v_2, v_3, v_{14}, and so on. Therefore, node v_4 is relatively more important. In addition, betweenness ranking method cannot tell the position difference and importance of edge nodes; if the betweenness of 12 nodes, including nodes v_8, v_{12}, v_{13}, v_{15}, and v_{16}, is 0, it means betweenness ranking method may fail in the evaluation of edge nodes. The trouble with the closeness ranking method is that the closeness value of each node is similar in the network, which cannot highlight the position difference of different nodes in the network. Moreover, the closeness ranking method tends to consider that nodes close to the center of the network are more important; it is effective for centralized networks, similar to a star network, but for more general networks, it may lead to the wrong judgment. To take nodes v_3 and v_{34}, for example, the closeness of node v_3 is relatively larger, while the degree of node v_{34} is the greatest in the network, and its betweenness is greater than node v_3, and, in the real social network, node v_{34}

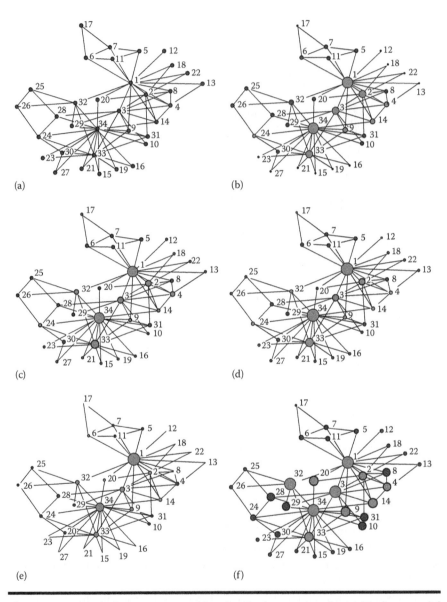

Figure 4.31 Node importance ranking for Zachary network: (a) Zachary social network, (b) top 10 nodes with maximum topological potential, (c) top 10 nodes with maximum degree, (d) top 10 nodes with maximum PageRank value, (e) top 10 nodes with maximum betweenness, and (f) top 10 nodes with maximum closeness.

Table 4.1 Top 10 Important Nodes for Zachary Network from Different Node Importance Ranking Algorithms

Node Importance Ranking	Degree Ranking Method		Betweenness Ranking Method		Closeness Ranking Method		PageRank Algorithm		Topological Potential Algorithm	
	Node No.	Degree	Node No.	Betweenness	Node No.	Closeness	Node No.	PageRank Value	Node No.	Potential Value
1	34	17	1	0.8238	1	0.0172	34	0.1009	34	0.2246
2	1	16	34	0.5724	3	0.0169	1	0.0970	1	0.2152
3	33	12	33	0.2734	34	0.0167	33	0.0717	33	0.1721
4	3	10	3	0.2704	32	0.0164	3	0.0571	3	0.1546
5	2	9	32	0.2603	9	0.0156	2	0.0529	2	0.1390
6	32	6	9	0.1053	14	0.0156	32	0.0372	32	0.1134
7	4	6	2	0.1015	33	0.0156	4	0.0359	4	0.1071
8	24	5	14	0.0863	20	0.0152	24	0.0315	9	0.1015
9	14	5	20	0.0611	2	0.0147	9	0.0298	14	0.1015
10	9	5	6	0.0564	4	0.0141	14	0.0295	24	0.0952

represents A. John as the core of a small group after the split; therefore, it may be concluded that node v_{34} should be more important than node v_3.

The PageRank and topological potential algorithms are degree-biased in nature, and consider that a node with a greater degree is more important. When node degrees are the same, for example, when the degrees of nodes v_4 and v_{32} are 6 and the degrees of nodes v_9, $v_{14,}$ and v_{24} are 5, a degree algorithm cannot distinguish the relative importance of these nodes, while PageRank and topological potential algorithms can make a finer importance distinction, both consider node v_{32} is more important than node v_4. However, the two algorithms are slightly different in ranking nodes v_9, v_{14}, and v_{24}; the PageRank algorithm considers that node v_{24} is the most important, while the topological potential algorithm considers that node v_9 is the most important. In an analysis of position difference between v_9 and v_{24} in the network, we find that node v_9 has larger betweenness, is closer to the network center, and is connected to the most important nodes v_1, v_3, v_{33}, and v_{34}; therefore, it is reasonable to believe that the node is relatively more important.

Theoretical analysis and experimental results show that when each node affects only 1-hop neighbors, a topological potential algorithm degrades to a degree algorithm; with the spreading influence of nodes, a topological potential algorithm will show an approximate linear correlation with the PageRank algorithm; and when the influence of the nodes spreads to 2-hop neighbors, 3-hop neighbors, and even the average diameter of the network, a topological potential algorithm will exhibit a centrality bias similar to closeness algorithm. By adjusting the parameters that reflect the influence of nodes, namely, impact factor σ, an importance ranking algorithm based on topological potential not only provides a better unified description framework for degree ranking method, closeness ranking method, and other common node importance ranking methods, but also produces node importance ranking more aligned with the network topology.

4.3.3 Network Community Discovery Based on Topological Potential

Community discovery in networks stems from sociological research and dates back to graph theory research. A traditional graph segmentation method always assumes that the network is decomposable and that the number of subgraphs to be segmented is specified by users. However, in the segmentation of a large-scale real network community, a community discovery method must find answers to two questions: whether community structure is included in a network, and how to effectively discover the community structure inherent in a network. So far, many community discovery methods have been proposed, most of whose basic ideas are to recursively merge or split the network based on node cohesion, and divide the network into a nested community structure. Typical representative methods include community discovery based on edge betweenness and modularity optimization.

It is not hard to find that a few routing nodes are often the only way for traffic to move between communities; if we can find the edge with the highest traffic, we can remove it to make a natural network division. Thus, Girvan and Newman introduced edge betweenness to measure network traffic and proposed a community discovery algorithm based on edge betweenness, referred to as the GN algorithm [16]. Its basic idea is to iteratively calculate the betweenness of each edge, and remove the edge with maximum betweenness until all edges are removed and each node becomes an independent community. Different from a graph segmentation method, the GN algorithm need not pre-specify the number of communities and can divide the network into any number of communities, but it cannot determine the optimal community structure. Thus, even for a random network that apparently has no community structure, GN algorithm can still produce a mandatory community division. In addition, the time complexity of the GN algorithm is high; if n is the number of network nodes, and m is the number of edges, the complexity is $O(nm^2)$.

For mandatory community divisions associated with the GN algorithm, Newman introduced modularity to evaluate the reasonableness of community division [17], and the basic idea is that in good community division, the connection probability of its internal nodes should be much greater than that of internal nodes in a random graph with the same degree sequence. Because the definition of modularity is independent of specific community structure, community discovery can be simplified to modularity optimization. As the number of possible divisions shows an exponential relation to the network scale, exhaustive discovery is infeasible for large-scale networks; thus, a variety of heuristic modularity optimization methods are introduced, such as greedy algorithm, extremal optimization algorithm, and simulated annealing algorithm.

Modularity optimization has become a basic community discovery method for complex networks and is widely used. However, there is an inherent resolution limit to the definition of modularity and community structures that are often found of a similar size. In fact, the size of the inherent community may be very uneven in a real network. Arenas et al. analyzed communities in e-mail networks, jazz and scientific collaboration networks, and so on, and found that the community sizes of these networks approximately followed the power-law distribution, namely, there are only a very small number of large communities in a network and most communities contain only a very few nodes. Apparently, for ubiquitous scale-free real networks, modularity optimization has obvious limitations.

Based on topological potential, we propose a novel community discovery algorithm [18]. This algorithm uses topological potential to describe the interaction between network nodes, considers the community as a high-potential area of the topological potential, and achieves network community division by finding high-potential areas separated by low-potential areas. The algorithm need not pre-specify parameters, such as the number of communities, can effectively reveal the community structure inherent in the network, reflects uncertain overlapping of nodes among communities, and has a better performance.

Given a network $G = (V, E)$, where $V = \{v_1, \ldots, v_n\}$ being node set, $E \subseteq V \times V$ being edge set, $|E| = m$ (m is the number of edges), community discovery based on topological potential can be described as follows:

1. *Topological potential attraction*: Given a local maximum-potential node v^*, $\forall v \in V$, if there is a node set of $\{v_0, v_1, \ldots, v_k\} \subset V$, let $v_0 = v$, $v_k = v^*$ and v_i be in the rise direction of the potential value of v^*, $0 < i < k$, v is said to be attracted by the topological potential of v^*.
2. *One representative community*: Given a local maximum-potential node v^*, if there is a subset of $C \subseteq V$, let $\forall v \in C$ and all be attracted by the topological potential of v^*, C is said to be a community represented by v^*, namely, the community only has one representative node.
3. *Multi-representatives community*: Given a set of local maximum-potential nodes $A \subset V$, if there is a subset of $C \subseteq V$, let (a) $\forall v \in C$, $\exists v^* \in A$, and v be attracted by the topological potential of v^*; (b) $\forall v^* \in A$, $\exists w^* \in A$, $w^* \neq v^*$, and distance $d\left(v^*, w^*\right) < \left\lfloor 3\sigma/\sqrt{2} \right\rfloor$, C is said to be a multi-representatives community represented by A.

Given a network, we should calculate its topological potential distribution, search the rise direction of maximum-potential value of each node for all local maximum-potential nodes, and determine the number of community divisions and the representative node for each community. There is only one local maximum-potential node in the one representative community, while there are multiple local maximum-potential nodes in the multi-representatives community, which form a ridge of obvious linear topological potential distribution, or a plateau of obvious plane topological potential distribution. We should combine the neighboring local maximum-potential nodes with distances less than $\left\lfloor 3\sigma/\sqrt{2} \right\rfloor$ into a multi-representatives set. In network division based on community representatives, $\forall v \in V$, if v is not a local maximum-potential node, then v may be attracted by only one community representative, or attracted by multiple community representatives at the same time. If v is attracted by only one community representative $C \subseteq V$, v is called an internal, or private, node of community C; if v is attracted by multiple community representatives, it is called an edge node.

Determination of community for an edge node is essentially an iterative expansion of intra-community nodes. Specifically, we assume that network G can be divided into t communities $C_1, \ldots, C_t$, and let $(a_{ij})_{n \times n}$ be an adjacency network matrix; for any edge node $v_i \in V$, we can introduce the following efficiency function Q to evaluate its allocation scheme and allocate it to the neighboring community with maximum efficiency. If the efficiencies of multiple allocation schemes are equal, then v_i is considered as an overlapping node, namely, the node is a fence-sitting node.

$$Q_{s=1,\ldots,t}\left(v_i\right) = \sum_{v_j \in V_s} a_{ij} - \sum_{v_k \notin V_s} a_{ik}$$

Community Discovery Algorithm Based on Topological Potential

Input: Network $G = (V, E)$ and impact factor σ, where, $V = \{v_1, \ldots, v_n\}$ and $|E| = m$.

Output: Communities $C_1, \ldots, C_t$.

Steps:

1. $[\varphi(v_1), \ldots, \varphi(v_n)] = Cal\_TopologicalPotential(G, \sigma)$; //Calculate node topological potential.
2. $V_{rep} = Searching\_MaxPotentialNodes(G, \varphi(v_1), \ldots, \varphi(v_n))$; //Use climbing method in rise direction of maximum-potential value to search local maximum-potential node, and determine community representatives set.
3. $[C_1, \ldots, C_t] = Community\_Detecting(G, V_{rep}, \varphi(v_1), \ldots, \varphi(v_n))$; //Detect the network community based on community representatives.
4. Output communities $C_1, \ldots, C_t$.

In an analysis of the above algorithm, the calculation overhead for topological potential approaches $O(n^2)$; in a search of all local maximum-potential nodes in the rise direction of maximum-potential value of each node, its time complexity is $O(m)$; in the network community division based on representatives set V_{rep}, if the number of representatives is $n_{rep} \ll n$, the time complexity for determination of respective communities for each node is $O(n_{rep} \times n) \sim O(n)$. Thus, the total time complexity of the algorithm depends on the calculation overhead for node topological potential, $O(n^2)$ in the worst case, and $O(m + n^{3/\gamma})$ in the best case.

We still use the Zachary social network to test the effectiveness of community discovery based on topological potential and compare it with GN algorithm, modularity optimization, and so on.

Figure 4.32a is the community structure inherent in a network; in a community division of the Zachary network using the community discovery algorithm based on topological potential, the parameters of potential entropy are first optimized, the value of σ is taken as 1.0203, and the topological potential distribution is as shown in Figure 4.32b, where the node's diameter is proportional to its topological potential value. Apparently, there are two local maximum-potential nodes, which, respectively, correspond to two core nodes v_1 and v_{34} in a real community structure. With the local maximum-potential node as a community representative node, the initial community, made up of internal nodes, is divided as shown in Figure 4.32c, including 17 nodes; after an iterative allocation of the 17 edge nodes, the final community structure is as shown in Figure 4.32d, where dotted lines are used for distinguishing, and v_{10} has the same connections with two communities, so can be

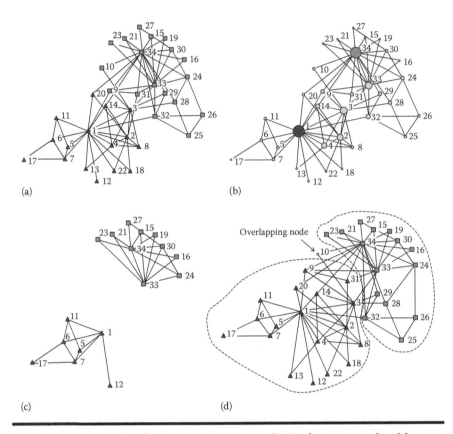

Figure 4.32 Analysis of community structure in Zachary network with community discovery based on topological potential: (a) real community structure, (b) topological potential distribution of nodes (σ = 1.0203), (c) initial community division, and (d) community division results.

considered an overlapping node. According to the topological potential values of the neighboring nodes, node v_{10} can be allocated to the community of the maximum-potential neighbor v_{34}. Thus, community discovery based on topological potential provides the same community structure as the Zachary network.

GN algorithm and modularity optimization are used in an analysis of community structure for the Zachary network. The GN algorithm needs to specify the number of communities to be divided; if the number of communities is 2, the GN algorithm may misclassify node v_3; modularity optimization does not need users to specify the number of communities to be divided, but it tends to produce a community structure with smaller granularity where the Zachary network is divided into five small communities, and node v_3 is misclassified. Compared with the previous two methods, the topological potential method does not need users to specify algorithm parameters, such as the number of communities, which can effectively reveal the community structure inherent in the Zachary network, such as the fence-sitting of nodes.

4.3.4 Hot Entries in Wikipedia Discovered with Topological Potential

Wikipedia was founded by Jimmy Wales and Larry Sanger in 2001. It is a model for Internet users to build a large-scale knowledge resource by means of free contributions and cooperation. As of 2012, *Wikipedia* contained more than 19 million entries, written in 285 languages [19]. Through an open editing and cooperation incentive mechanism, *Wikipedia* reflects the swarm wisdom of different knowledge backgrounds, while retaining modification histories as outdated or incorrect interpretations of entries are updated by other users to show the evolution of the entries. It plays an important role for humans to realize a knowledge-sharing mechanism in the network era, and for knowledge to approach truth.

Based on the Chinese *Wikipedia*, we downloaded entry data in a computer field from 2003 to 2009 as a typical study case. We took the entry as a node and hyperlinks to other entries in its explanation as sides. Without considering the direction, we constructed a network topology graph for the link relationships of computer field entries in Chinese *Wikipedia* from 2003 to 2009, as shown in Figure 4.33. Based on this graph, we analyzed the macro statistical properties and the dynamic evolution laws of the entry link network topology in this section, and mined the hot entries, based on the topological potential [20,21].

Because Chinese *Wikipedia* was founded in October 2002, the network was still small-scale in 2003, containing only 183 nodes, which included six entries as isolated nodes—namely, calculator, quaternion, discrete amount, Fermat's last theorem, computational mathematics, and mathematical constant—while the remaining 177 nodes form one large connected sub-graph with 914 sides, with nodes represented by circles, as shown in Figure 4.33a. With more and more volunteers participating in the editing, the number of *Wikipedia* entries has been growing exponentially. In 2005, the network developed to contain 1,802 nodes and 10,733 sides. To show the growth of nodes, the entries in 2003 are marked yellow, and the newly added nodes from 2003 to 2005 are marked red, as shown in Figure 4.33b. In 2007, the entry relationship network in *Wikipedia* scaled up further, to 1,885 nodes and 19,436 sides. Although the growth rate slowed, the editing relationship between entries increased substantially, as shown in Figure 4.33c. The nodes for 2003 are marked yellow, those produced between 2003 and 2005 are labeled green and those produced between 2005 and 2007 are marked red. In the growth process, a large number of wrongly defined entries were deleted and the correct contents have gradually stabilized. There was little change in the total number of nodes, but the

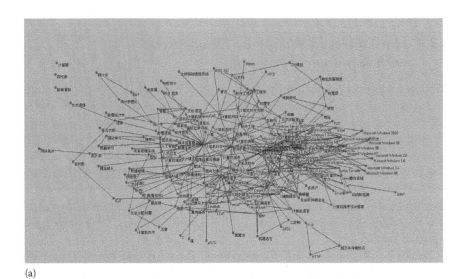

(a)

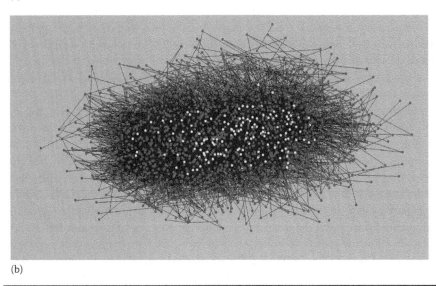

(b)

Figure 4.33 Topological evolution of an entry relation network in the field of computers in Chinese *Wikipedia* from 2003 to 2009: (a) year 2003, (b) year 2005.
(*Continued*)

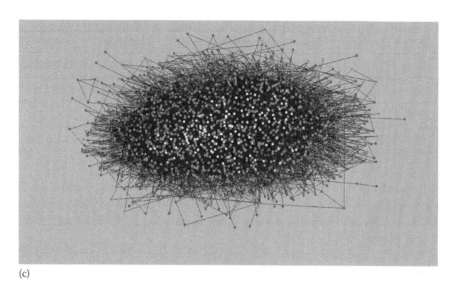

(c)

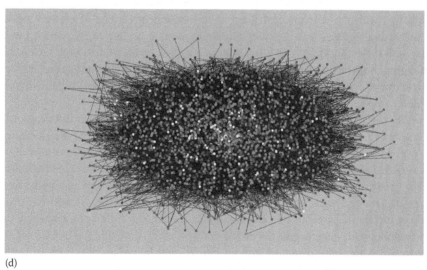

(d)

Figure 4.33 (*Continued*) Topological evolution of an entry relation network in the field of computers in Chinese *Wikipedia* from 2003 to 2009: (c) year 2007, and (d) year 2009.

Table 4.2   Attributes of the Connectivity between Computer Entries in *Wikipedia* 2003–2009

Years	Nodes	Number of Sides	Diameter	Average Path Length	Average Clustering Coefficient	Average Degree
2003	177	914	7	3.390537	0.476888	10.327684
2005	1802	10,733	8	3.592348	0.376265	11.912320
2007	1885	19,436	7	3.190981	0.363681	20.621751
2009	2796	39,828	7	3.173849	0.424217	28.489270

link relationship between nodes increased significantly, exhibiting that the quality of information in *Wikipedia* has improved steadily. By the end of 2009, computer entries in *Wikipedia* experienced a large growth, boasting 2,796 nodes and 39,828 sides, as shown in Figure 4.33d.

Through a statistical analysis of the maximal connected sub-graph in *Wikipedia* between 2003 and 2009, we get the basic statistics for computer entries network by year, as shown in Table 4.2. By analyzing the network's statistical parameters we find that, with the network scaled up, the average degree value of network nodes is increasing significantly. By 2009, the average value is close to 29, which shows that with the evolution of computer science a large number of new word entries are emerging, but there is no significant increase in the average shortest path length or the network diameter; there is a short distance between two nodes in the network, which complies with small-world characteristics.

We counted the form of their degree distribution in double logarithmic coordinates, and fit its power-law index γ, as shown in Figure 4.34. We found that, although the network scale is largely different in each year, its degree distribution shows obvious power-law characteristics. In particular, the tail of topological form of 2009 obviously presents as heavy with the probability of $P(k)$ decreasing. Very few nodes have large degrees and most nodes have small degrees, which further proves that there exists a preferential attachment phenomenon in the long evolution process, where the entry nodes with large degree are linked with priority. Although *Wikipedia*'s entries are free to share, for interaction without barriers and by groups of people, from the perspective of topological characteristics, there is still a close relationship between few entries with large degree and enormous entries with general degree. This kind of connection makes the network structure of *Wikipedia* entries flatter, indicating that entry knowledge is conveyed faster, more directly, and more efficiently.

Through modeling the entry link relationship network in Chinese *Wikipedia* for the years 2003–2009, we can use the method for ranking node importance based on topological potential, as proposed in Section 4.3.2, to mine the hot entries and reveal development trends in computer science in different years. Figure 4.35 shows

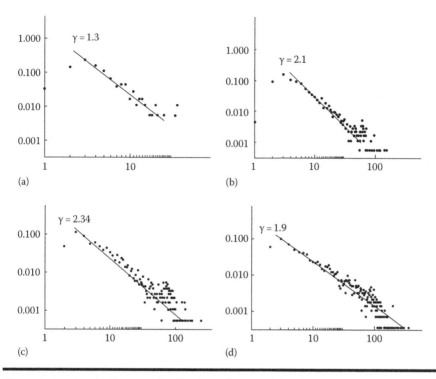

Figure 4.34 Node degrees distribution of computer-related entries relation network in Chinese *Wikipedia* by year: (a) year 2003, (b) year 2005, (c) year 2007, and (d) year 2009.

the topological graph for popular *Wikipedia* entries from 2003 to 2009. Because of large number of nodes, and to mark the changing trend in hot entries in the network, the diameter of each node in the graph is positively proportional to its topological potential value; the greater the topological potential value, the greater the diameter of the corresponding node, indicating it has a more important position in the network.

By calculating the topological potential value, we can also rank the importance of entries by year. Table 4.3 lists the top 10 important entries and their potential values with the maximum-potential values by year. We can see that with the evolution of computer-related entries in *Wikipedia*, people's attention to programming language has changed significantly. For example, the early C language used for basic function programming ranked in the top 8 in 2003, then the cross-platform Java stabilized in first position, while the dynamic programming language PHP, Python for network interaction, and then the .NET framework all appeared in hot entries, reflecting how programming language is evolving in the trend for simpler employment and more agile development. Java's topological potential value keeps increasing, maintaining its first ranking, which fully reflects the significant advantages of the language.

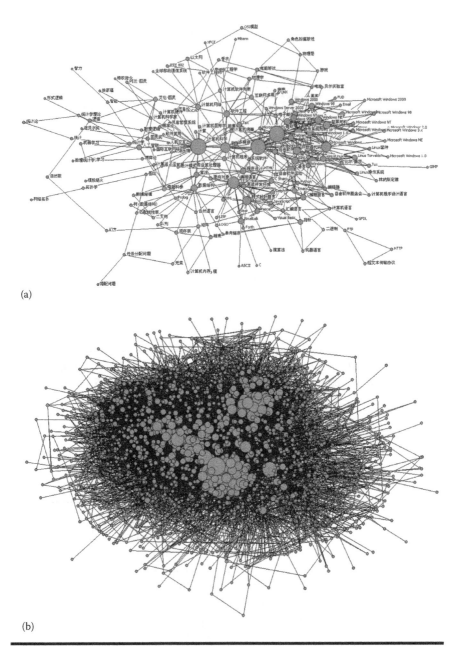

(a)

(b)

Figure 4.35 Topological structure of computer-related entries in Chinese *Wikipedia* **by year: (a) year 2003, (b) year 2005.** **(*Continued*)**

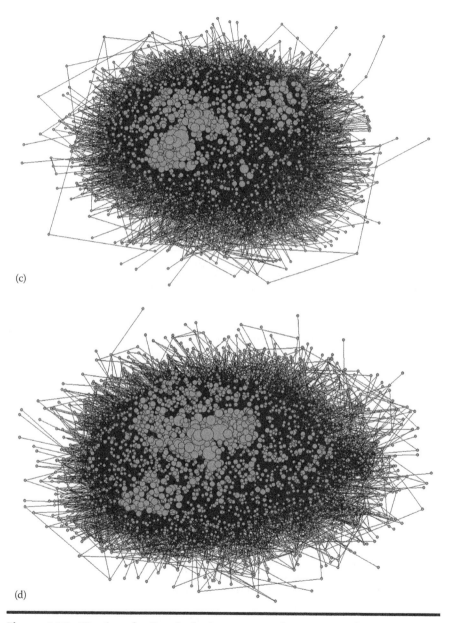

(c)

(d)

Figure 4.35 (*Continued*) Topological structure of computer-related entries in Chinese *Wikipedia* by year: (c) year 2007, and (d) year 2009.

Table 4.3   Top 10 Hot Entries Obtained by Using Topological Potential for 2003–2009

2003		*2005*	
Entries	*Topological Potential Value*	*Entries*	*Topological Potential Value*
Computer Science	34.25645	Java	50.63085
Operating system	33.04369	Linux	48.03578
Microsoft	30.06516	Computer Science	39.92616
Programming language	26.50462	Operating system	37.97985
Software	25.83673	Microsoft Windows	36.03354
Linux	21.03906	PHP	32.14093
Microsoft Windows	19.82223	Perl	32.14093
C language	19.08877	C language	32.14093
Software engineering	18.41681	Programming language	31.81654
List of Operating System	17.46227	JavaScript	30.19462
2007		*2009*	
Entries	*Topological Potential Value*	*Entries*	*Topological Potential Value*
Java	82.09617	Java	123.293
Operating system	59.71362	Microsoft Windows	102.208
VBScript	58.0917	Operating system	96.04471
Linux	57.44293	Linux	91.17894
Microsoft Windows	55.49662	Software development	84.69124
Visual Basic .NET	51.92839	Jscript	81.4474
Microsoft	51.60401	VBScript	80.79863
PHP	50.95524	Operating system	78.20355
Python	47.71139	.NET framework	75.28409
Perl	47.38701	Internet Explorer	73.66217

Physics studies the most general laws of nature and physical methods are typical representatives of scientific method. The cognitive field method extends cognition of the objective world in physics to the cognition of the subjective world by human beings, using data fields to describe complex interactions and interactions between primitive, chaotic, and unstructured data objects, so as to realize the self-organized clustering and simplified induction of large-scale data sets. The topological potential method can extend the physical field method to the understanding and recognition of the complex natures of complex systems, such as their structure, function, and dynamic behavior, as a result opening up new thinking in the research of complex networks and complexity science.

References

1. D. Li, W. Gan, and L. Liu, Artificial intelligence and cognitive physics. In: *Proceedings of the 10th Annual Conference of Chinese Association of Artificial Intelligence (CAAI-10)*, Guangzhou, China, 2003.
2. W. Gan, Study on clustering problem for data mining foundations, PhD thesis, PLA University of Science and Technology, Nanjing, China, 2003.
3. W. Gan, D. Li, and J. Wang, An hierarchical clustering method based on data fields, *Chinese Journal of Electronics*, 34(2), 258–262, 2006.
4. W. Gan, D. Li, and J. Wang, Dynamic clustering based on data field. In: *The 11th International Fuzzy Systems Association World Congress on Fuzzy Logic, Soft Computing and Computational Intelligence*, Beijing, China, 2005.
5. S. Wang, W. Gan, D. Li et al., Data field for hierarchical clustering, *International Journal of Data Warehousing and Mining*, 7(4), 43–63, 2011.
6. S. Guha, R. Rasogi, and K. Shim, CURE: An efficient clustering algorithm for large databases. In: *Proceedings of the 1998 ACM SIGMOD International Conference on Management of Data*, Seattle, WA, 1998.
7. R. Duda and P. Hart, *Pattern Classification and Scene Analysis*, John Wiley & Sons, New York, 1973.
8. T. Zhang, R. Ramakrishnman, and M. Linvy, BIRCH: An efficient method for very large databases. In: *Proceedings of ACM SIGMOD International Conference on Management of Data*, Montreal, Quebec, Canada, 1996.
9. X. Dai, W. Gan, and D. Li, Study of image data mining based on data field, *Computer Engineering and Applications*, 26, 41–44 2004.
10. M. J. Lyons, S. Akamatsu, M. Kamachi et al., Coding facial expressions with Gabor wavelets. In: *Proceedings of the Third IEEE International Conference on Automatic Face and Gesture Recognition*, Nara, Japan, 1998.
11. D. J. Watts and S. H. Strogatz, Collective dynamics of small world networks, *Nature*, 393(6684), 440–442, 1998.
12. A. L. Barabási and E. Bonabeau, Scale-free networks, *Scientific American*, 288(5), 50–59, 2003.
13. W. Gan, Data field method and its application in network data mining, Postdoctoral report, Tsinghua University, Beijing, China, 2007.

14. N. He, W. Gan, and D. Li, Evaluate nodes importance in the network using data field theory. In: *The 2007 International Conference on Convergence Information Technology*, Gyeongju, Korea, 2007.

15. W. W. Zachary, An information flow model for conflict and fission in small groups, *Journal of Anthropological Research*, 33(4), 452–473, 1997.

16. M. Girvan and M. E. J. Newman, Community structure in social and biological networks, *Proceedings of the National Academy of Sciences of the United States of America*, 99(12), 7821–7826, 2002.

17. M. E. J. Newman, Modularity and community structure in networks, *Proceedings of the National Academy of Sciences of the United States of America*, 103(23), 8577–8582, 2006.

18. W. Gan, N. He, D. Li et al., Community discovery method in networks based on topological potential, *Journal of Software*, 20(8), 2241–2254, 2009.

19. Wikimedia Foundation, 2012-5-7. http://meta.wikimedia.org/wiki/Main_Page.

20. Y. Han, Network topology measurement and multi granularity mining method, Master's degree thesis, Beihang University, Beijing, China, 2010.

21. H. Zhang, G. Zhang, Y. Ma et al., Group interests and their correlations mining based on Wikipedia, *Chinese Journal of Computers*, 34(11), 2234–2242, 2011.

Chapter 5

Reasoning and Control of Qualitative Knowledge

Human intelligence with uncertainty reflects human adaptability to an uncertain environment. People have been utilizing some classic experiments to simulate this kind of uncertain intelligence: chess playing between human and computer, soccer matches between robots, and controlling a multilevel inverted pendulum.

Over the past 60 years artificial intelligence has made great progress in proving symbolic theorems and in logical reasoning, resulting in the defeat of Lee Sedol by AlphaGo. In the control of an inverted pendulum, which is a classic research topic in automatic control, balance and robustness are regarded as an effective way to evaluate intelligent control. An intelligent vehicle is a kind of complex wheeled robot. With the development in recent years of cloud computing, network navigation, and intelligent transportation technology, intelligent vehicles have received a lot of attention and achieved a boom in development. The robot soccer match also reflects the progress in uncertain processing related to computer vision, planning, cooperation, and control, all important in the research on AI with uncertainty.

In this chapter, we are going to study reasoning and control with qualitative knowledge represented by a cloud model rather than through a precise mathematical model, and implementing stable control by qualitative rule controller, reflecting the control experience expressed by human natural language.

5.1 Cloud Reasoning

Knowledge is generally those concepts that are the result of continuous abstraction and communication, and their interrelations. In the control field, rules such as "perception–action" are usually adopted to express the logic cause-result relation. The precondition, such as perception, may be composed of one or multiple conditions, and the postcondition, such as the action, represents concrete control behavior. The concepts of both precondition and postcondition can be uncertain.

5.1.1 Using a Cloud Model to Construct Qualitative Rules

The cloud-based qualitative rule generator is composed of a precondition cloud generator and a postcondition cloud generator.

According to the mathematical properties of cloud mentioned in Chapter 2, we can see that the qualitative concept C can be expressed by the Gaussian cloud $C(Ex, En, He)$ in the universal set U. The distribution of the cloud drops in U has expectation Ex and variance $En^2 + He^2$.

Suppose there is a rule

$$\text{If } A \text{ then } B$$

where A and B correspond to the concepts C_1 and C_2 in the universal sets U_1 and U_2, respectively.

Let a be a certain point in U_1, the distribution of certainty degree of a that belongs to C_1 can be generated by a cloud generator, called a precondition cloud generator, as shown in Figure 5.1a. Let μ, $\mu \in [0, 1]$ be a certainty degree, the distribution of cloud drops that satisfies this certainty degree on the concept B in U_2 can be generated by a cloud generator, called a postcondition cloud generator, as shown in Figure 5.1b.

U_1 and U_2 can be either one-dimensional or multidimensional. Generally, if the precondition/postcondition cloud generator is multidimensional, it can be descended to be several one-dimensional precondition cloud generators.

The algorithm of the one-dimensional precondition cloud generator is illustrated in Section 5.1.1.1.

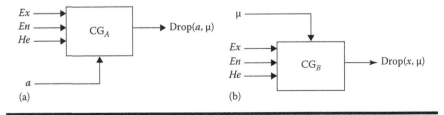

Figure 5.1 One-dimensional (a) precondition and (b) postcondition cloud generators.

5.1.1.1 One-Dimensional Precondition Cloud Generator [1]

Input: The digital characteristics of the qualitative concept C_1 (expectation Ex, entropy En, hyper-entropy He, a specific value a)

Output: The cloud drop corresponding to the specific value a and its certainty degree μ.

Steps:

```
BEGIN
    En' = NORM(En, He²);
```
$$\mu = e^{\frac{-(a-Ex)^2}{2(En')^2}};$$
```
    OUTPUT drop (a, μ);
END
```

The specific value a and the certainty degree μ, which is generated by the precondition cloud generator, construct the joint distribution (a, μ), as shown in Figure 5.2a. All cloud drops are distributed in the line $x = a$.

The algorithm of the postcondition cloud generator is provided as follows.

5.1.1.2 Postcondition Cloud Generator [1]

Input: The digital characteristics (Ex, En, He) of the qualitative concept B, certainty degree μ

Output: The cloud drop b with certainty degree μ.

Steps:

```
BEGIN
    En' = NORM(En, He²);
    b = Ex ± En'√-2lnμ;
    OUTPUT drop (b, μ);
END
```

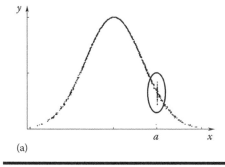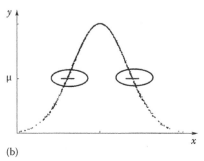

Figure 5.2 Joint distribution of cloud drops: (a) cloud drops distribution of precondition cloud generator and (b) cloud drops distribution of postcondition cloud generator.

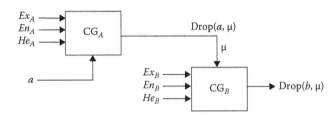

Figure 5.3   Single-condition-single-rule generator.

The specific certainty degree μ and the drop b produced by the postcondition cloud generator construct the joint distribution (b, μ) as shown in Figure 5.2b. All the drops are in the same line $y = \mu$. According to Section 2.4, the drops satisfy the two normal distributions with expectation $EX = Ex + \sqrt{-2\ln\mu}En$ and variance $DX = -2He^2\ln\mu$, and expectation $EX = Ex - \sqrt{-2\ln\mu}En$ and variance $DX = -2He^2\ln\mu$, respectively.

The single-condition-single-rule can be expressed by

$$\text{If } A \text{ then } B$$

where A and B are qualitative concepts. For example, in the rule "if the altitude is high, then the population density is low," A represents the concept of "high altitude" and B represents the concept of "low population density." If we combine one precondition cloud generator with one postcondition cloud generator, as shown in Figure 5.3, we obtain a single conditional rule, called single-condition-single-rule generator.

5.1.1.3 Single-Condition-Single-Rule Generator

Input: The qualitative concept $A(Ex_A, En_A, He_A)$ of the precondition; The qualitative concept $B(Ex_B, En_B, He_B)$ of the postcondition; A specific value a in the universal set U_1 of the precondition.

Output: The cloud drop b in the universal set U_2 of the postcondition, and its certainty degree μ.

Steps:

```
BEGIN
    // Generate a normally random number En'ₐ with expectation
      Enₐ and deviation He²ₐ
```
$$En'_A = NORM\left(En_A, He_A^2\right);$$
$$\mu = e^{-\frac{(a - Ex_A)^2}{2En_A'^2}};$$
```
    // Generate a normally random number En'_B with expectation
      En_B and deviation He²_B.
```

```
En'ᵦ = NORM (Enᵦ, He²ᵦ);
//If the input activates the rising edge of the
   precondition, then the postcondition is also at its
   rising condition, and vice versa.
if a < Ex then
    b = Exᵦ - √(-2ln(μ))En'ᵦ
else
    b = Exᵦ + √(-2ln(μ))En'ᵦ;
OUTPUT(b, μ);
END
```

In the single-condition-single-rule generator, if a specific input a in the universal set U_1 of the precondition activates CG_A, it will randomly generate a certainty degree μ. This value reveals the activation strength on the qualitative rule by a, and it acts as the input for the postcondition cloud generator to generate drop(b, μ) randomly. If a activates the rising edge of the precondition, then the output b by the rule generator corresponds to the rising edge of the postcondition, and vice versa.

Uncertainty is contained in this algorithm. For a certain input a in U_1, it cannot output a certain b. The certainty degree μ randomly generated by CG_A transits the uncertainty in U_1 to U_2. CG_B outputs random cloud drop(b, μ) under the control of μ, so drop b is also uncertain. In this way, uncertainty is transited in the reasoning process by rule generators.

To research a multi-condition-single-rule generator, we first discuss the construction of a double-condition-single-rule generator.

The double-condition-single-rule can be expressed by

$$\text{If } A_1, A_2 \text{ then } B$$

For example, the qualitative rule "if temperature is high and pressure is high, then speed is fast" reflects the relation between three qualitative concepts in the domains of temperature, pressure, and speed, that is, A_1 = "high," A_2 = "high," and B = "fast."

Two precondition cloud generators and one postcondition cloud generator can be connected, according to Figure 5.4, so as to construct a double-condition rule, called a "double-condition-single-rule generator."

The precondition includes the "and" relation of two qualitative concepts, and this relation is hidden in the expression by natural language. A_1, A_2, and B correspond to the concepts in U_{A_1}, U_{A_2}, and U_B, respectively. The degree of "and" between μ_1 generated by CG_{A_1} and μ_2 generated by CG_{A_2} are not stated clearly in logic. Hence we can introduce a new concept, that is, "soft and," to implement the generation of μ from μ_1 and μ_2, and construct a double-condition-single-rule generator.

We regard "soft and" as a qualitative concept expressed by a two-dimensional Gaussian cloud $C(1, Enx, Hex, 1, Eny, Hey)$. The universal sets of the two dimensions correspond to the value domains of the certainty degrees μ_1 and μ_2, and they are both in [0, 1]. The expectation of the drops (x, y) is (1, 1). At this point, the

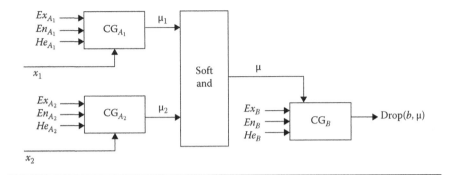

Figure 5.4   Double-condition-single-rule generator.

certainty degree μ is strictly 1, and so the "soft and" is equivalent to logic "and." The certainty degrees of the cloud drops in other positions are less than 1. This shows the uncertainty of "and," which is the special property of "soft and." The farther the distance from the cloud drop to the expectation point, the lower the certainty degree μ [2].

Enx, Eny, Hex, and *Hey* can be used as the adjustment parameters for the degree of "soft and." When *Enx* = *Eny* = 0 and *Hex* = *Hey* = 0, "soft and" will be degenerated to logic "and." In the definition of "soft and," the output cloud drop and its joint distribution (*x, y,* μ) are meaningful only with *x* and *y* in [0, 1]. Thus, the cloud graph of the joint distribution looks like a quarter of a hill, as illustrated in Figure 5.5.

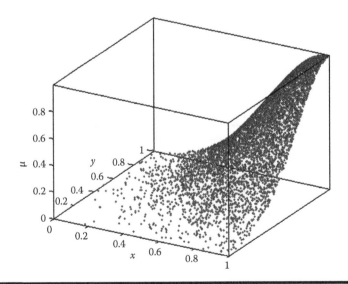

Figure 5.5   The cloud graph of "soft and."

Compared to the selection of min$\{\mu_1, \mu_2\}$ from the fuzzy set as the strength of postcondition activation, it is more adaptable to realize the quantitative transformation of "soft and" with a cloud model.

The multi-condition-single-rule generator can be constructed by expanding the method utilized in the double-condition-single-rule generator.

The single-condition-single-rule and multi-condition-single-rule previously generated can be stored in the rule database for further reasoning and control of the qualitative knowledge.

5.1.2 Generation of Rule Sets

In the practice of reasoning and qualitative control, it is difficult to give an accurate mathematical model for the controlled object. However, there usually exists a large amount of data containing precise output, activated by precise input, based on practical manipulation. Such information may construct classical cases, if typical enough, such as

"If the temperature is 50°F and the pressure is 40 Pa, then the speed is 40 r/s";
"If the temperature is 55°F and the pressure is 60 Pa, then the speed is 60 r/s";
"If the temperature is 70°F and the pressure is 80 Pa, then the speed is 90 r/s";
"If the temperature is 90°F and the pressure is 90 Pa, then the speed is 100 r/s";
$$\vdots$$
$$\vdots$$

The word "classical" emphasizes that these given cases come out of a long-term accumulation of experience and are examples that must be obeyed.

The number of these cases and their data distribution reflect key control points in reasoning and control, and represent turning points in the nonlinear relationship between input and output. It might be true that it is difficult to express these nonlinear relationships in terms of mathematical functions, but people can realize the reasoning by comparing these classical cases, known as case-based reasoning [3]. Its main strategy is that with a given input, it searches for and matches a case in the case database, and generates the output. Generally, a complete match with a classical case is not possible, so modification is required for the output. The shortcomings in this approach are that the control of some key points cannot cover all possibilities. To address this problem, we can first abstract precise cases for qualitative concepts expressed by linguistic values, and then construct qualitative rule sets to provide a reasoning mechanism. In these qualitative rules, each linguistic value can be expressed by three digital characteristics of the cloud model, regardless of their semantic meanings.

Once these qualitative rules are obtained, the output with uncertainty can be generated through a reasoning mechanism with the new input.

5.2 Cloud Control

This section compares the cloud-based qualitative control method, the classical fuzzy control method, and the probability control method to illustrate the qualitative control mechanism of the cloud-based method. All three methods are utilized to solve the uncertainty problems in the control process.

5.2.1 Mechanism of Cloud Control

To facilitate discussion, we first provide a classical case used by the fuzzy control and probability control methods [4]. This is about control of motor speed based on temperature variation. There is a rule database containing five qualitative rules, as shown in Table 5.1.

Note that, in these five qualitative rules, the variables in the precondition and the postcondition are natural linguistic values rather than precise numbers. What would the output of motor speed be with a precise temperature input, such as 68°F?

Fuzzy control constructs the rule database with fuzzy rules. If the input condition is determined, it first calculates the membership degree of all the qualitative concepts in the precondition, and then regards this as the strength to activate the postcondition. If there are multiple rules activated simultaneously, the reasoning engine will be enabled to solve it and to give a certain output.

Table 5.1   Rule Base

	Temperature		Motor Speed
	Cold		Stop
	Cool		Slow
If	Just right	Then	Medium
	Warm		Fast
	Hot		Blast

In this example, the control mechanism of fuzzy control goes:

1. Assume that the membership functions of the five qualitative concepts in the precondition are triangle shaped functions in the temperature domain, as shown in Figure 5.6. They can be expressed in mathematics as

$$
\mu_{cold}(x) = \begin{cases} 1 & x \in [30,40] \\ 1 - \dfrac{x-40}{50-40} & x \in [40,50] \\ 0 & \text{otherwise} \end{cases}
$$

$$
\mu_{cool}(x) = \begin{cases} \dfrac{x-45}{55-45} & x \in [45,55] \\ 1 - \dfrac{x-55}{65-55} & x \in [55,65] \\ 0 & \text{otherwise} \end{cases}
$$

$$
\mu_{justright}(x) = \begin{cases} \dfrac{x-60}{65-60} & x \in [60,65] \\ 1 - \dfrac{x-65}{70-65} & x \in [65,70] \\ 0 & \text{otherwise} \end{cases}
$$

$$
\mu_{warm}(x) = \begin{cases} \dfrac{x-65}{75-65} & x \in [65,75] \\ 1 - \dfrac{x-75}{85-75} & x \in [75,85] \\ 0 & \text{otherwise} \end{cases}
$$

$$
\mu_{hot}(x) = \begin{cases} \dfrac{x-80}{90-80} & x \in [80,90] \\ 1 & x \in [90,100] \\ 0 & \text{otherwise} \end{cases}
$$

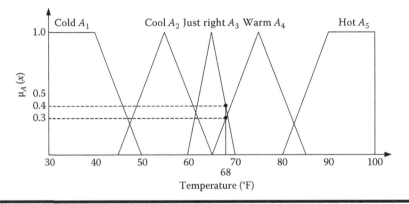

Figure 5.6 Membership functions of the qualitative concepts in the precondition.

The membership functions of the five qualitative concepts in the postcondition are also defined as triangle shaped functions in the speed domain, as shown in Figure 5.7. They are expressed in mathematics as follows:

$$\mu_{stop}(z) = \begin{cases} 1 - \dfrac{z}{30} & z \in [0,30] \\ 0 & \text{otherwise} \end{cases}$$

$$\mu_{slow}(z) = \begin{cases} \dfrac{z-10}{30-10} & z \in [10,30] \\ 1 - \dfrac{z-30}{50-30} & z \in [30,50] \\ 0 & \text{otherwise} \end{cases}$$

$$\mu_{medium}(z) = \begin{cases} \dfrac{z-40}{50-40} & z \in [40,50] \\ 1 - \dfrac{z-50}{60-50} & z \in [50,60] \\ 0 & \text{otherwise} \end{cases}$$

$$\mu_{fast}(z) = \begin{cases} \dfrac{z-50}{70-50} & z \in [50,70] \\ 1 - \dfrac{z-70}{90-70} & z \in [70,90] \\ 0 & \text{otherwise} \end{cases}$$

$$\mu_{blast}(z) = \begin{cases} \dfrac{z-70}{100-70} & z \in [70,100] \\ 0 & \text{otherwise} \end{cases}$$

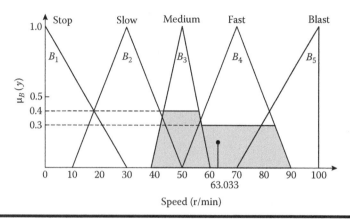

Figure 5.7 The Mamdani fuzzy control method.

If there is a precise input t in the temperature domain, the method calculates the membership degrees of the qualitative concepts in the precondition for each of the five fuzzy rules.

2. If there is only one positive membership degree μ_i, the ith rule is activated with strength μ_i, and the program goes to step 3. If there is more than one positive membership degree, μ_i, and μ_{i+1}, the ith and $i + 1$th rules are activated with strengths μ_i and μ_{i+1}, respectively, and the program goes to step 4.

3. If there is only one activated rule, the program directly cuts the membership function of the postcondition by μ_i, and obtains the precise output according to the criteria that if it is on the rising/dropping edge, the corresponding rising/dropping edge will be activated.

4. If there are multiple activated rules, the Mamdani fuzzy control method (product-sum-gravity methods) [5] will be utilized. The multi-activation strength will be applied to cut the postcondition of related rules to obtain the overlapping of several trapeziums. Thereafter, the center of area (COA) will be calculated to generate precise output.

In this example, when a precise temperature, $t = 68°F$, is input to the system, the third and fourth rules will be activated, with activation strengths 0.4 and 0.3, as illustrated in Figure 5.6.

Thereafter, the membership functions of the postcondition are cut by the two activation strengths, as shown in Figure 5.7. By means of the Mamdani fuzzy control method, the COA of the shadow area, which is the sum of the reaction area of the two rules, will be calculated for precise output of the reasoning:

$$COA = \frac{\int_{-\infty}^{\infty} y \times \mu_B(y)\,dy}{\int_{-\infty}^{\infty} \mu_B(y)\,dy} = \frac{942.35}{14.95} = 63.033 \text{ r/min}$$

In other words, when the input temperature is 68°F, the motor speed given by the fuzzy controller will be 63.033 r/min.

The probability control method constructs the rule database with rules with probability. Its control mechanism is as follows:

1. Define the conditional probability distribution functions of the five qualitative concepts in the precondition as

$$
P_1[cold|x] = \begin{cases} 1 & x \in [30,45] \\ \dfrac{(50-x)}{5} & x \in [45,50] \\ 0 & \text{otherwise} \end{cases}
$$

$$
P_2[cool|x] = \begin{cases} \dfrac{(x-45)}{5} & x \in [45,50] \\ 1 & x \in [50,60] \\ \dfrac{(65-x)}{5} & x \in [60,65] \\ 0 & \text{otherwise} \end{cases}
$$

$$
P_3\big[just\ right|x\big] = \begin{cases} \dfrac{(x-60)}{5} & x \in [60,65] \\ \dfrac{(70-x)}{5} & x \in [65,70] \\ 0 & \text{otherwise} \end{cases}
$$

$$
P_4[warm|x] = \begin{cases} \dfrac{(x-65)}{5} & x \in [65,50] \\ 1 & x \in [70,80] \\ \dfrac{(85-x)}{5} & x \in [80,85] \\ 0 & \text{otherwise} \end{cases}
$$

$$
P_5[hot|x] = \begin{cases} \dfrac{(x-80)}{5} & x \in [80,85] \\ 1 & x \in [85,100] \\ 0 & \text{otherwise} \end{cases}
$$

From this definition, it can be seen that the sum of the probabilities with any point belonging to one of the five concepts should be 1. The probability density functions of the precondition are illustrated in Figure 5.8.

The probability density functions (Figure 5.9) of the five qualitative concepts in the postcondition can be defined with

$$f_1(z) = \begin{cases} \dfrac{(30-z)}{450} & z \in [0,30] \\ 0 & \text{otherwise} \end{cases}$$

$$f_2(z) = \begin{cases} \dfrac{(z-10)}{400} & z \in [10,30] \\ \dfrac{(50-z)}{400} & z \in [30,50] \\ 0 & \text{otherwise} \end{cases}$$

$$f_3(z) = \begin{cases} \dfrac{(z-40)}{100} & z \in [40,50] \\ \dfrac{(60-z)}{100} & z \in [50,60] \\ 0 & \text{otherwise} \end{cases}$$

$$f_4(z) = \begin{cases} \dfrac{(z-50)}{400} & z \in [50,70] \\ \dfrac{(90-z)}{400} & z \in [70,90] \\ 0 & \text{otherwise} \end{cases}$$

$$f_5(z) = \begin{cases} \dfrac{(z-70)}{450} & z \in [70,100] \\ 0 & \text{otherwise} \end{cases}$$

2. Once a temperature t is input to the system, the five conditional probability density functions of the qualitative concepts are calculated for each probability rule to obtain p_1, p_2, p_3, p_4, p_5
3. The output, that is, the motor speed, is a random value z with probability density function

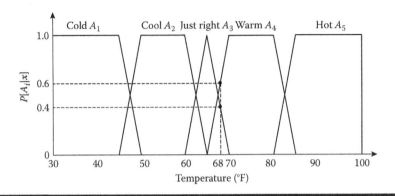

Figure 5.8 Probability density functions of the qualitative concepts in the precondition.

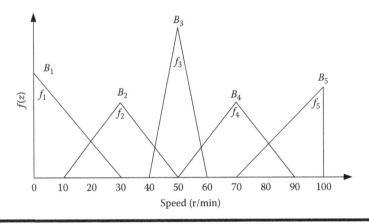

Figure 5.9 Probability density functions of the qualitative concepts in the postcondition.

$$f(z) = p_1 \times f_1(z) + p_2 \times f_2(z) + p_3 \times f_3(z) + p_4 \times f_4(z) + p_5 \times f_5(z)$$

When a precise input $t = 68°F$ is input to the system, all the other conditional probabilities are 0, except $p["just\ right"|68°F] = 0.4$ and $p["warm"|68°F] = 0.6$. Because the expectations of $f_3(z)$ and $f_4(z)$ are 50 and 70 r/min, respectively, the expectation of the output motor speed is

$$\text{MEAN}(z) = 0 + 0 + 0.4 \times 50 + 0.6 \times 70 + 0 = 62 \text{ r/min}$$

For the same input temperature $t = 68°F$, the motor speed obtained by the fuzzy control method is constantly 63.033 r/min, while the expectation of speed given by the probability control method is 62 r/min.

After this rough introduction to fuzzy control and probability control, it is necessary to describe the reasoning mechanism of the cloud-based rule generator in detail.

1. For easy comparison with the aforementioned two methods, the five cloud qualitative concepts in the temperature domain are expressed as

$$A_1 = \begin{cases} 1 & x \in [30,40] \\ C(30,20/3,0.05) & \text{otherwise} \end{cases}$$

$$A_2 = C(55,10/3,0.05)$$

$$A_3 = C(65,5/3,0.05)$$

$$A_4 = C(75,10/3,0.05)$$

$$A_5 = \begin{cases} C(90,10/3,0.05) & \text{otherwise} \\ 1 & x \in [90,100] \end{cases}$$

The joint distribution between the cloud drops of the five qualitative concepts in the precondition and their certainty degrees are shown in Figure 5.10.

Similarly, the clouds for the five qualitative concepts in the postcondition are expressed as

$$B_1 = C(0,10,0.05)$$

$$B_2 = C(30,20/3,0.05)$$

$$B_3 = C(50,10/3,0.05)$$

$$B_4 = C(70,20/3,0.05)$$

$$B_5 = C(100,10,0.05)$$

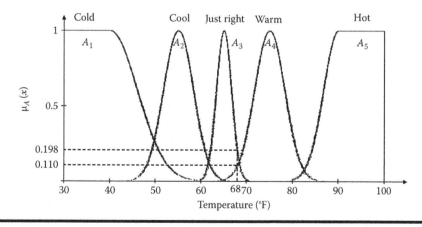

Figure 5.10   Cloud graph of the qualitative concepts in the precondition.

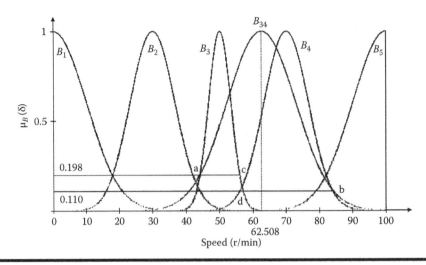

Figure 5.11 Cloud graph of the qualitative concepts in the postcondition and virtual cloud.

The joint distribution between the cloud drops of the five qualitative concepts in the postcondition and their certainty degrees are shown in Figure 5.11.

2. When there is a precise input t in the temperature domain, we calculate its certainty degrees to the five qualitative concepts in the preconditions of each of the five cloud rules.

3. If there is only one positive certainty degree μ_i, then the ith rule will be activated with activation strength of μ_i, and the output will be generated directly by the single-rule-generator.

4. If there is more than one positive certainty degree, we assume they are μ_i and μ_{i+1}. Then the ith and $i + 1$th rules will be activated with strengths of μ_i and μ_{i+1}, respectively. The output will be constructed using the virtual cloud.

In this example, when $t = 68°F$ is the input to the system, the third and fourth rules are activated with strengths 0.198 and 0.110, as shown in Figure 5.10. Four cloud drops will be generated after the activation of the postcondition by these two degrees of uncertainty, as shown in Figure 5.11. The two cloud drops at the two sides are selected to construct a virtual concept $C_{B_{34}}$, which is also expressed by the cloud. The formation of the virtual cloud is as follows.

Let $B_{34}(Ex, En, He)$ be a virtual cloud with the same shape. It covers the two drops $a(x_1, \mu_1)$ and $b(x_2, \mu_2)$. Set $He = 0$ temporarily, as we have only the position of the two drops at this moment. Using a geometrical method the expectation and the entropy of the virtual cloud B_{34} can be calculated as

$$Ex = \frac{x_1\sqrt{-2\ln(\mu_2)} + x_2\sqrt{-2\ln(\mu_1)}}{\sqrt{-2\ln(\mu_1)} + \sqrt{-2\ln(\mu_2)}}$$

$$En = \frac{x_2 - x_1}{\sqrt{-2\ln(\mu_1)} + \sqrt{-2\ln(\mu_2)}}$$

If there are more cloud drops $(x_1, \mu_1), (x_2, \mu_2), \ldots, (x_n, \mu_n)$, the backward cloud algorithm can produce the expectation, entropy, and the hyper-entropy of the virtual cloud.

In this example, the expectation of the virtual concept $C_{B_{34}}$ is $Ex = 62.508$ r/min, which is the output of the motor speed.

A similar method can be utilized to obtain the motor speed for different input temperatures.

In comparing these three control mechanisms, we notice that with the fuzzy control method, not only is a certain membership function required, but the motor speed is constant for the same input temperature. By contrast, with probability control and cloud-based control, the outputs are uncertain every time, but the same overall variation trend is maintained as with the fuzzy control method, as shown in Figure 5.12. The cloud-base method has also avoided the strict condition required by the probability method, that is, the sum of the probability density functions of "stop," "slow," "medium," "fast," and "blast" should be 1. The description for these

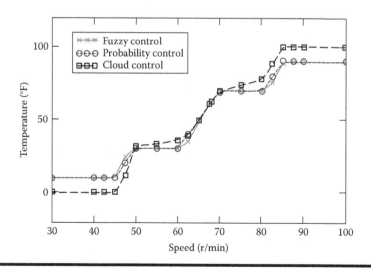

Figure 5.12 **Corresponding relation between input and output by different control methods.**

linguistic values is similar to that for the temperature values, "cold," "cool," "just right," "warm," and "hot." As expressed in terms of expectation, entropy, and hyperentropy, they are more understandable.

5.2.2 Theoretical Explanation of the Mamdani Fuzzy Control Method

The Mamdani method is applied ubiquitously in fuzzy control and it adopts the computation method of "product-sum-gravity," when multiple rules are activated simultaneously. No theoretical explanation of the Mamdani method is provided in the literature, but we shall attempt one based on cloud-based control.

We can still take Figure 5.7 as an example. When two rules are activated simultaneously, the two adjacent activated concepts in postcondition can be utilized to construct a new virtual concept whose central value corresponds to the output value. It can be proved that this central value is the gravity value obtained by the Mamdani control method [6]. For simplicity we substitute the Gaussian cloud with the simple triangular cloud, as shown in Figure 5.13, B_{34} is the new virtual concept formed after the activation of B_3 and B_4.

Because

$$Area\,|_{\Delta B_3 Q_1 Q_2} = Area\,|_{\Delta B_{34} Q_1 Q_2}$$

$$Area\,|_{\Delta B_4 Q_3 Q_4} = Area\,|_{\Delta B_{34} Q_3 Q_4}$$

$$Area\,|_{\Delta Q_1 K_1 Z_1} \approx Area\,|_{\Delta Q_2 T_1 Z_2}$$

$$Area\,|_{\Delta Q_3 Z_3 T_1} \approx Area\,|_{\Delta Q_4 Z_4 K_2}$$

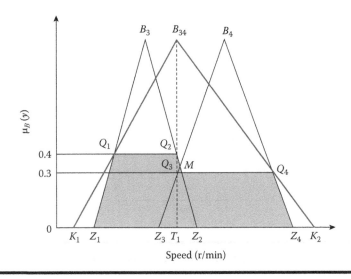

Figure 5.13 Explanation of the Mamdani method by the cloud-based method.

So

$$COA\big|_{polygon Q_1 Z_1 Z_4 Q_4 M Q_2} \approx COA\big|_{\triangle B_{34} K_1 K_2}$$

The sum of the areas of B_3 and B_4 approximates the area of B_{34}, so the horizontal coordinate of the shadow part is the same as the expectation of the virtual concept B_{34}.

When the Gaussian cloud is applied to represent the qualitative concept in the rule, a similar conclusion can be drawn. In this way, we illustrate the theoretical explanation of the Mamdani control method using the virtual cloud concept. The Mamdani method contains the virtual concept generated by the concepts in the two activated rules.

5.3 Uncertainty Control in Inverted Pendulum

Reasoning and control based on qualitative knowledge have been widely applied not only in expert systems, decision support, and similar areas, but also successfully utilized in the field of automatic control.

What creatures, nature, and man-made systems have most in common is that they all realize system stabilization by feedback. In the over 3000 years of scientific and technological development, humans have designed numerous systems based on feedback control. The progress of civilization has been marked by numerous examples, such as dams in ancient times, the swing clock and the telescope in medieval times, steam machines in the industrial revolution, followed by modern airplanes, automobiles, telephones, analog computers, radars, satellites, missiles, digital computers, and space shuttles. All these well-known inventions have directly promoted and boosted the technology of automatic control.

5.3.1 Inverted Pendulum System and Its Control

In the development of automatic control theory and technology, the correctness of a certain theory is usually verified through the control of an object by a controller designed according to this theory. The inverted pendulum is one that has had perpetual popularity, and it can be studied with several control theories and methods, such as PID (proportional-integral-derivative) control, adaptive control, state feedback control, neural network control, fuzzy control, and so on. When a brand new control method is announced that cannot be proven on a theoretical level, its correctness and physical applicability can be verified with the inverted pendulum.

What is an inverted pendulum then? We know that there are circus performers who hold up long bars on their heads, with heavy objects on the end. Such poles can be abstractly considered as unstable pendulums that are inverted, that is, their center of gravity is higher than the supporting point. Their stability can be maintained by an intuitive human performer, and by qualitative control methods.

For such a complex, time-varying, strong coupling relation it is very difficult to use mathematical methods to quantify the precise characterization of such non-linear biological modeling and control systems. If there are changes to the external work environment of the pendulum, traditional control methods are even more unfeasible. For decades, people have compared pendulum research to a "pearl in the crown" pursued by automatic control researchers.

Research on the inverted pendulum is of great engineering value. Making robots stand and walk is similar to a double-link inverted pendulum system. It has been more than 60 years since the first robot was created in the United States; however, the key technology in the robot—walking control—remains a challenging problem. A slight trembling in the reconnaissance satellite will have a huge effect on image quality, so automatic stabilization is required to eliminate such trembling. A flexible (multistage) rocket was invented to prevent the breakup of a single-stage rocket when turning, and consequently the attitude control of this system can also be studied via the multistage inverted pendulum.

In conclusion, research on the inverted pendulum is of significant practical value. In the past few decades, hundreds of papers on the inverted pendulum have been published by scholars from the United States, Japan, Hong Kong, Canada, Sweden, Norway, and more. There have been inspiring results on single, double, and triple one-dimensional inverted pendulums, and on single- and double-link inverted pendulums on a plane. Some research output has also been harvested by national universities and institutes in China. Presently, research of the inverted pendulum has become a common educational tool in automation teaching.

5.3.2 Qualitative Control Mechanisms for Single-Link/Double-Link Inverted Pendulums

Let us start with the single-link inverted pendulum, as shown in Figure 5.14. The inverted pendulum L_1 is equivalent to the bar used in the circus, and the car is

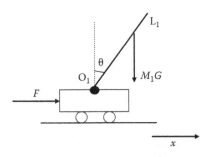

Figure 5.14   Diagram of a single-link inverted pendulum.

equivalent to the performer. The pendulum and the car are connected by a non-frictional rotational potentiometer. The external force F on the car, and the moved distance x, are equivalent to the manipulation behavior of the performer. θ is the angle from the vertical line to the pendulum, which is defined as positive if the pendulum is on the right. This angle can be measured by the potentiometer. The origin of the moved distance is in the middle of the track. x is defined as positive if it is on the right of the origin.

The control objective of a single-link inverted pendulum is to stabilize the inverted pendulum, that is, $\theta \approx 0$ and $x \approx 0$, by the appropriate controlling forces on the track with a finite length, even after disturbance.

It can be seen that the force transition between the pendulum and the car is a "coupling" relation, where external force F on the car is transited to the inverted pendulum and makes it stable. Combined with human intuition, it is easy to qualitatively analyze the single-link inverted pendulum system as a qualitative physical model [7,8].

If there is no external force F, once the pendulum tilts to the left, the gravitational moment on the pendulum will speed up the tilting to the left, and the car will move to the right. Once the pendulum tilts to the right, it will speed up the movement and the car will move to the left.

If external force F is to the right, the car will tend to move faster to the right, and the pendulum will tend to tilt to the left. If F is to the left, the car will tend to move to the left, and the pendulum will tilt to the right.

The double-link inverted pendulum system has one more pendulum added to the single-link inverted pendulum. They are connected with one rotational potentiometer of the same model to measure the second angle. This system is more difficult to control than the single one. However, their control mechanisms are similar, with the only difference found in the compulsory force that is exerted by the car on the first pendulum by F' through the first one on the second pendulum. In this way, F' is an indirect controlling force on the second pendulum, so the control on the second pendulum by the car is also indirect.

Figure 5.15 shows the double-link inverted pendulum. θ_2 is the angle from the extended line of the first pendulum to the second pendulum, and clockwise direction is positive. The definition of θ_2 here is different from that in other literatures (θ_2 is the angle from the vertical line to the second pendulum in the literature), so that the measured value can be directly revealed to facilitate analysis.

The double-link inverted pendulum system aims to control the car appropriately so as to stabilize the double-link pendulum without divergent oscillation and place the car at $x \approx 0$, even after disturbance.

Qualitative analysis of this system is easily associated with two cases. The first one is that of a circus performer who steps on to board with a rolling ball underneath. He has to swing his upper body to keep his balance, so the lower part can be considered as the first pendulum and the upper part as the second one. However, the actuating force on the upper part is exerted by the waist, which equals a motor

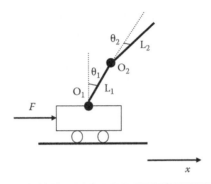

Figure 5.15   Diagram of a double-link inverted pendulum.

between the first and second pendulums. Thus it is different from the inverted pendulum system discussed. The other case is to erect two wooden bars connected by a rotational link. The controller must concentrate his attention on the upper pendulum while manipulating the lower bar, which is extremely difficult. This case is similar to the inverted pendulum system discussed. If the connection between the lower bar and the palm is not guaranteed, delay in controlling the force will be longer. Combined with a human's practical controlling experience, it is not difficult to work out the qualitative physical model of a double-link inverted pendulum system.

Let us assume no external force F is exerted on the system. Under the function of the gravitational moment of the pendulum, if the upper pendulum L_2 tilts to the right, O_2 will move to the left, and thus L_1 will tilt to the left by a certain angle, so the car will move to the right. All the motion will be opposite if the initial tilting is opposite.

If an external force F is exerted to the left, L_1 will tilt to the right, and point O_2 will move to the right; consequently L_1 will move to the left. Similarly, if the force is directed to the right, the whole process will go in the opposite direction.

In summary, once L_2 tilts a little, the car will exert an indirect force F' at point O_2 through L_1, which will make L_2 return to a dynamically stable position. This is the basic mechanism for the car to control the double-link inverted pendulum.

5.3.3 Cloud Control Strategy for a Triple-Link Inverted Pendulum

The triple-link inverted pendulum is a system with an additional pendulum added to the double-link inverted pendulum, as shown in Figure 5.16. The rotational potentiometer to measure the third angle is the same as that of the other two. It aims to keep the pendulums stable through the motion of the car, even after disturbance.

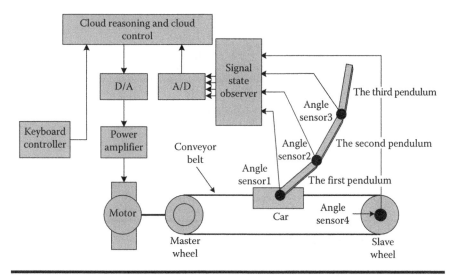

Figure 5.16   Experimental block diagram of a triple-link inverted pendulum.

As a matter of fact, a physical multilink inverted pendulum is a multivariate, nonlinear, fast-reactive, and unstable system with a lot of uncertainties in an actual environment. There is no accurate mathematical model to quantitatively describe it. However, experts in these fields may well achieve control with control rules extracted from their long experience and represented by intuitive natural language. Natural languages play a very important role in representing the uncertainty in human control and reasoning [9–11].

5.3.3.1 Qualitative Analysis of the Triple-Link Inverted Pendulum System

Figure 5.17 is the diagram of a triple-link inverted pendulum system, which is obviously a strongly coupled system. There are eight state variables to represent the system. The control objective can be expressed visually by $\theta_1 \rightarrow 0$, $\theta_2 \rightarrow 0$, $\theta_3 \rightarrow 0$, $x \rightarrow 0$.

x: the position of the car on the track
$\dot{x}$: the velocity of the car
θ_1: the tilting angle of the first pendulum with respect to the vertical direction
θ_2: the tilting angle of the second pendulum with respect to the first pendulum
θ_3: the tilting angle of the third pendulum with respect to the second pendulum
$\dot{\theta}_1, \dot{\theta}_2, \dot{\theta}_3$: the angular velocity of the first, second, and third pendulums

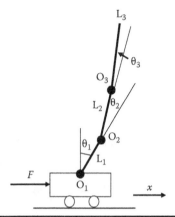

Figure 5.17   Diagram of a triple-link inverted pendulum.

After a simple analysis of the triple-link inverted pendulum system, we can draw the following qualitative conclusions:

1. If θ_1, θ_2, and θ_3 are initialized as zero and the gravitational moment of L_1, L_2, and L_3 are neglected, the force F to the right will make the first pendulum tilt to the left, that is, the position of O_1 will move to the right, O_2 to the left, and O_3 to the right. Similarly, if the force is directed to the left, the whole process will be in the opposite direction.

2. θ_1 and θ_2 are initialized as zero and F is neglected. If the third pendulum tilts to the right, its motion will be accelerated by its gravitational moment, and consequently the second pendulum will tilt to the left and the first one to the right, that is, O_3 will move to the left, O_2 to the right, O_1 to the left, and the car to the left. The interaction from the upper pendulum to the lower one, and the gravitational function of the lower pendulum itself, will accelerate the variation. Similarly, if the third pendulum tilts to the left, the process will be in the opposite direction.

3. θ_1 and θ_3 are initialized as zero and F is neglected. If the second pendulum tilts to the right, the gravity of the third pendulum, together with the gravitational moment of the first pendulum, will increase the motion of the second one to the right. The first and third pendulums will tilt to the left, that is, O_2 will move to the left, O_3 to the right, O_1 to the right, and the car to the right. The gravitational interaction from the upper pendulum to the lower one, together with the gravitational moment of the lower pendulum itself, will accelerate the variation. Similarly, if the second pendulum tilts to the left, the process will be in the opposite direction.

4. θ_2 and θ_3 are initialized as zero and F is neglected. If the first pendulum tilts to the right, the gravities of the upper pendulums, together with the gravitational moment of itself, will further the motion to the right. Meanwhile, the third pendulum will move to the right, the second to the left, that is, O_3 will move to the left, O_2 to the right, O_1 to the left, and the car to the left. The gravitational interaction from the upper pendulum to the lower one, together with the gravitational moment of the lower pendulum itself, will accelerate the variation. Similarly, if the first pendulum tilts to the left, the process will be in opposite direction.

Obviously, force F is directly exerted on the car and transited to L_3 through L_1 and L_2 by means of coupling. Hence, to achieve stabilization, first, $\theta_3 = 0°$ should be satisfied, then the second pendulum should be stabilized ($\theta_2 = 0°$), followed by the stabilization of the lowest pendulum, that is, $\theta_1 = 0°$, and finally the stabilization of the car ($x = 0$) will be considered. As a result, the third pendulum can be regarded as the most significant. The parameters of its rule generator should be most important, and those of the second and first pendulums, and then car will be in descending order.

We can also decompose the triple-link inverted pendulum system into a single inverted pendulum with a two-stage inverted pendulum so as to convert the control of x and $\dot{x}$ of single-link inverted pendulum to the control of O_3 of the double-link inverted pendulum. The control of O_3 can be achieved by the car to promote the double-link pendulum.

Finally, it is necessary to discuss the relation between differential control $D(\dot{\theta}_1, \dot{\theta}_2, \dot{\theta}_3, \dot{x})$ and proportional control $P(\theta_1, \theta_2, \theta_3, x)$. In the system, proportional control has the advantage of immediately reacting once an error occurs; however, if the pendulum is self-stable, there will be a static error. The differential control, on one hand, is sensitive to the variation trend of the error, and the reactive response can be speeded up to reduce the overshooting and increase the stability; on the other hand, it is also sensitive to noise, which will reduce system stability. Hence, it is necessary to select appropriate proportional control to reduce the error and choose proper differential control to enhance the stability. The weight relation between them is more important than that within each control.

After analysis, we get the following order of importance of each triple-inverted pendulum system semaphores

$$\theta_3, \dot{\theta}_3, \theta_2, \dot{\theta}_2, \theta_1, \dot{\theta}_1, x, \dot{x}$$

Based on this, we design a cloud controller that utilizes the cloud-based rule generator to control the signals.

5.3.3.2 The Cloud Controller of the Triple-Link Inverted Pendulum System

The cloud controller (Figure 5.18) mainly employs the cloud generator to control various signals in the triple-link inverted pendulum system, whose design involves the following aspects:

1. Determine the input and output variables of the cloud controller.
2. Design the control rules of the controller, including the number and the content of the rules.
3. Select the cloud types for expression of the preconditions and postconditions of the rules.
4. Select the universal sets of the input and output variables of the cloud controller, and determine their parameters, such as the quantitative factor, the three digital characteristics of the cloud, and so on.
5. Program the algorithm for the cloud controller.
6. Select an appropriate sampling time for the cloud controller algorithm.

For those controlled systems that cannot be described in precise mathematical formalization, experts can construct a qualitative rules set based on natural language to reflect human reasoning ability about the system's nonlinear and variable granularity, to control these systems successfully. The cloud model just acts as a qualitative–quantitative transformation model, using digital technology to realize computer control through uncertainty reasoning.

Some terms in the conditional statements that describe the control rules, such as "large," "small," "high," and so on, have a certain degree of uncertainty. We can use the cloud model to represent the values of these words and the rule generator to represent the rules. By using a computer program to implement these control rules, the computer acts as a controller.

In real applications, as people are accustomed to dividing events into two hierarchies, we can describe the input/output states of the cloud controller with the words "large" and "small." Considering the zero state of the variable, there are five qualitative concepts altogether, that is, negatively large, negatively small, zero, positively small, and positively large. Based on this, five concepts can be defined for each input/output variable in the construction of the cloud rule generator. The concrete

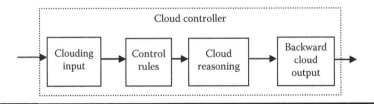

Figure 5.18 Theoretical block diagram of the cloud controller.

control rules of the triple-link inverted pendulum are illustrated in Table 5.2, and the parameter set of the qualitative concepts in the rules is shown in Table 5.3.

In Table 5.3, the two qualitative concepts, "positively large" and "negatively large," are expressed by half-Gaussian clouds, and the other three by Gaussian clouds. The parameters given here are not the angle or displacement values, but the number corresponding to the sampling value after the conversion of the A/D plate.

It can be seen from Figure 5.19, the control flow chart of the triple-link inverted pendulum, that five rule controllers on eight signals are utilized, including the three tilting angles $(\theta_3, \theta_2, \theta_1)$, their differentials $(\dot{\theta}_3, \dot{\theta}_2, \dot{\theta}_1)$, the displacement of the car x, and the speed of the car $\dot{x}$. According to the significance order of these signals, we have to prioritize some signals and control them once they exceed the range. In the experiment, five rule controllers control the pendulum sequentially, according to their priorities by means of "soft zero," which is a Gaussian random number with expectation zero and entropy En. In the control, it will vary each time. According to the judgment conditions in the five rule controllers, there will be related "soft zeros"—$SZ_{\theta_3}, SZ_{\theta_2}, SZ_{\theta_1}, SZ_x,$ $SZ_{\dot{\theta}_3}, SZ_{\dot{\theta}_2}, SZ_{\dot{\theta}_1}, SZ_{\dot{x}}$—the entropies of which effectively determine the probabilities of entering into a related rule controller. The range and probability of the "soft zero" will vary according to different entropies. These "soft zeros" may be classified into "big zero," "medium zero," and "small zero" based on their values. Consequently, the transition between the rule controllers can be realized by the values of the "soft zeros."

We have implemented effective control over a triple-link inverted pendulum via a cloud control method. The eight signals of the triple-link inverted pendulums are illustrated in Figure 5.20. Figure 5.21 shows the graph of the triple-link inverted pendulum demonstrated in IFAC'99 in July 1999 in Beijing [12]. In the demonstration, the system showed excellent stability, even in cases where a bunch of flowers or knocking disturbance were added to the system.

5.3.4 Balancing Patterns of the Inverted Pendulum

In the inverted pendulum system, the rotational potentiometers are joints. Although they are frictionless, the difference in agility of the joints will result in different balancing patterns of the system under control. The discussion of the balancing patterns will benefit other research, such as the multistage flexible rocket.

The association degree ρ between the pendulums, which measures the coupling between the two pendulums, is defined as

$$\rho_n = 1 - \frac{|\theta_n|}{90°}$$

where
  θ_n is the measured angle of the nth pendulum ($n = 1, 2, \ldots$)
  ρ_1 is the association degree of the first pendulum and the car
  ρ_n is the association degree of the $n − 1$th and nth pendulum [13]

Table 5.2   Control Rules of Triple-Link Inverted Pendulum

The rules set RS(θ_3) of the tilting angle of the third pendulum θ_3			
	Tilting Angle θ_3		*Motor Output Force F*
	Positively large		Positively large
	Positively small		Positively small
If	Zero	Then	Zero
	Negatively small		Negatively small
	Negatively large		Negatively large
The rules set RS(θ_2) of the tilting angle of the second pendulum θ_2			
	Tilting Angle θ_2		*Motor Output Force F*
	Positively large		Negatively large
	Positively small		Negatively small
If	Zero	Then	Zero
	Negatively small		Positively small
	negative large		Positively large
The rules set RS(θ_1) of the tilting angle of the second pendulum θ_1			
	Tilting Angle θ_1		*Motor Output Force F*
	Positively large		Positively large
	Positively small		Positively small
If	Zero	Then	Zero
	Negatively small		Negatively small
	Negatively large		Negatively large

(Continued)

Table 5.2 (*Continued*)   Control Rules of Triple-Link Inverted Pendulum

	The rules set RS(x) of the car displacement x			
	Car Displacement x			Motor Output Force F
If	Positively large	Then		Negatively large
	Positively small			Negatively small
	Zero			Zero
	Negatively small			Positively small
	Negatively large			Positively large

	The rules set $RS(\dot{\theta}_3)$ of the angular velocity $\dot{\theta}_3$ of the third pendulum			
	Angular Velocity $\dot{\theta}_3$			Motor Output Force F
If	Positively large	Then		Positively large
	Positively small			Positively small
	Zero			Zero
	Negatively small			Negatively small
	Negatively large			Negatively large

	The rules set $RS(\dot{\theta}_2)$ of the angular velocity $\dot{\theta}_2$ of the second pendulum			
	Angular Velocity $\dot{\theta}_2$			Motor Output Force F
If	Positively large	Then		Negatively large
	Positively small			Negatively small
	Zero			Zero
	Negatively small			Positively small
	Negatively large			Positively large

(*Continued*)

Table 5.2 (*Continued*)   Control Rules of Triple-Link Inverted Pendulum

	The rules set $RS(\dot{\theta}_1)$ of the angular velocity $\dot{\theta}_1$ of the first pendulum			
	Angular Velocity $\dot{\theta}_1$			*Motor Output Force F*
	Positively large			Positively large
	Positively small			Positively small
If	Zero		Then	Zero
	Negatively small			Negatively small
	Negatively large			Negatively large
	The rules set $RS(\dot{x})$ of the car velocity $\dot{x}$			
	Car Velocity $\dot{x}$			*Motor Output Force F*
	Positively large			Negatively large
	Positively small			Negatively small
If	Zero		Then	Zero
	Negatively small			Positively small
	Negatively large			Positively large

When ρ_n approaches 1, the association degree of the corresponding pendulums becomes large, which means that the coupling between the two adjacent pendulums is so tight that they move like one pendulum. When ρ_n approaches 0, the association degree is small, resulting in weak coupling so that the swing amplitude of the lower pendulum is apparently greater than that of the upper one.

5.3.4.1 Balancing Patterns of the Single-Link Inverted Pendulum

In the example of raising the bar by a circus performer, the performer concentrates on the bar on his shoulder. He can either move his shoulder once tilting occurs or not move until the tilting increases to some extent. From such a phenomenon we can come up with two kinds of balancing pattern for a single-link inverted pendulum system. From a different perspective, there is only one association degree ρ_1, which has the values of "large" and "small," so the single-link

Table 5.3 Parameter Set of Qualitative Concepts in the Rule Generators for a Triple-Link Inverted Pendulum System

Precondition	Negatively Large	Negatively Small	Zero	Positively Small	Positively Large
θ_1	(−150, 31, 0.059)	(−57, 19, 0.04)	(0, 11.4, 0.0256)	(57, 19, 0.04)	(150, 31, 0.059)
$\dot{\theta}_1$	(−550, 30, 0.03)	(−209, 69.3, 0.02)	(0, 42, 0.0128)	(209, 69.3, 0.02)	(550, 30, 0.03)
θ_2	(−500, 103.5, 0.044)	(−190, 63, 0.03)	(0, 38, 0.0192)	(190, 63, 0.03)	(500, 103.5, 0.044)
$\dot{\theta}_2$	(−700, 145, 0.06)	(−266, 88, 0.033)	(0, 53.2, 0.0224)	(266, 88, 0.033)	(700, 145, 0.06)
θ_3	(−800, 166, 0.044)	(−304, 101, 0.03)	(0, 60.8, 0.0192)	(304, 101, 0.03)	(800, 166, 0.044)
$\dot{\theta}_3$	(−850, 176, 0.06)	(−323, 107, 0.033)	(0, 640, 0.0224)	(323, 107, 0.033)	(850, 176, 0.06)
x	(−265, 54.0, 0.023)	(−100, 33.4, 0.015)	(0, 20.2, 0.01)	(100, 33.4, 0.015)	(265, 54.9, 0.023)
$\dot{x}$	(−100, 20.7, 0.023)	(−38, 12.6, 0.015)	(0, 7.6, 0.01)	(38, 12.6, 0.015)	(100, 20.7, 0.023)
F	(−100, 20.7, 0.023)	(−38, 12.6, 0.015)	(0, 7.6, 0.01)	(38, 12.6, 0.015)	(100, 20.7, 0.023)

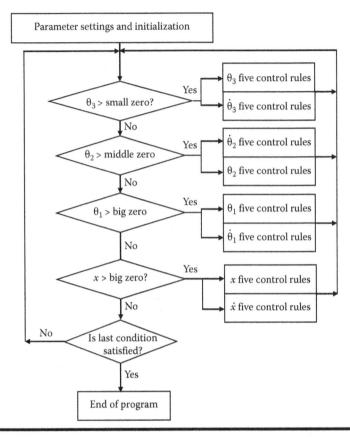

Figure 5.19   Control flow chart of the triple-link inverted pendulum.

inverted pendulum system has two balancing patterns, that is, large ρ_1 and small ρ_1.

1. *Balancing pattern* 1: Large ρ_1, as shown in Figure 5.22. At this moment, the association between the first pendulum and the car is large. The pendulum stands up on the car, and the car moves slowly on the track with small displacement.
2. *Balancing pattern* 2: Small ρ_1, as shown in Figure 5.23. At this moment, the association between the first pendulum and the car is small. The car moves back and forth regularly on the track with large displacement.

5.3.4.2 Balancing Patterns of the Double-Link Inverted Pendulum

There are two association degrees, that is, ρ_1 and ρ_2, in a double-link inverted pendulum system. If each of the two association degrees has the value of "large" and

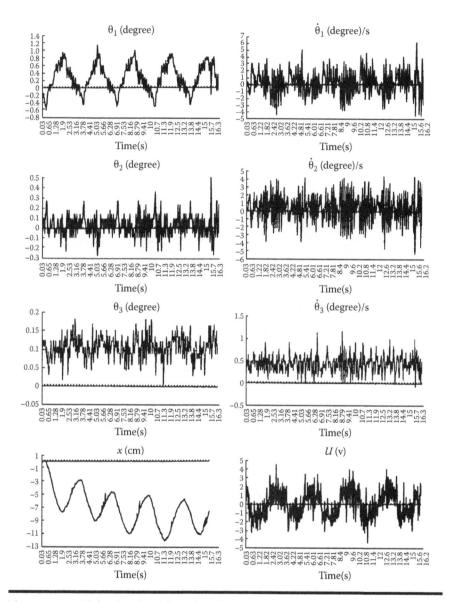

Figure 5.20   Eight signals of the triple-link inverted pendulum.

"small," there will be four combinations, that is, large ρ_1 with large ρ_2, small ρ_1 with relatively large ρ_2, small ρ_1 with small ρ_2, and large ρ_1 with relatively small ρ_2.

The fourth state, "large ρ_1 with relatively small ρ_2," is unstable, as shown in Figure 5.24. When the second pendulum tilts to the right, θ_2 and $\dot{\theta}_2$ will increase. At this moment, O_2 will move to the right to try to keep the pendulum stable, resulting in the left motion of the car. Consequently, θ_1 and $\dot{\theta}_1$ will increase positively.

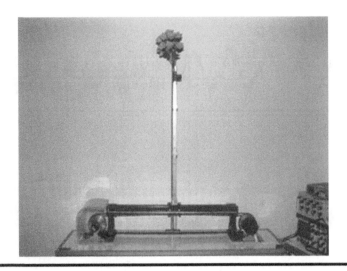

Figure 5.21 Robust experiment of the triple-link inverted pendulum.

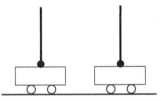

Figure 5.22 Balancing pattern 1 of the single-link inverted pendulum (large ρ_1).

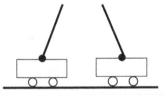

Figure 5.23 Balancing pattern 2 of the single-link inverted pendulum (small ρ_1).

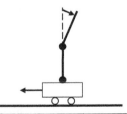

Figure 5.24 Unstable double-link inverted pendulum (large ρ_1 with relatively small ρ_2).

To guarantee the stability of the first pendulum, the motor should be controlled for a larger force to the right. In this process, the car has to adjust control of O_1 before it is able to control O_2, so the motion of the car cannot be stabilized. As a result, there is no balancing pattern with large ρ_1 and relatively small ρ_2. It can be concluded that under any circumstance, ρ_2 should be greater than ρ_1. In an n link inverted pendulum system, there is no state of large ρ_{n-1} with relatively small ρ_n, large ρ_{n-2} with relatively small ρ_{n-1}, ..., and large ρ_1 with relatively small ρ_2. Thus, there are altogether three classical balancing patterns of the double-link inverted pendulum system.

1. *Balancing pattern 1*: Large ρ_1 with large ρ_2, as shown in Figure 5.25. The association degrees between the first and the second pendulums are relatively large, so they stand up on the car, which moves slowly back and forth with small displacement.
2. *Balancing pattern 2*: Small ρ_1 with relatively large ρ_2, as shown in Figure 5.26. In this case, the association degree between the first and the second pendulums is large, while that between the first pendulum and the car is relatively small. As a result, the system looks like one pendulum tilting back and forth, apparently to keep balance.
3. *Balancing pattern 3*: Small ρ_1 with small ρ_2, as shown in Figure 5.27. In this case, the association degrees between the two pendulums and between the first pendulum and the car are small; however, because ρ_2 is greater than

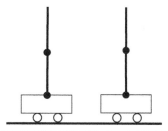

Figure 5.25   Balancing pattern 1 of the double-link inverted pendulum (large ρ_1 with large ρ_2).

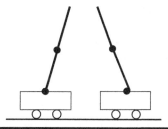

Figure 5.26   Balancing pattern 2 of the double-link inverted pendulum (small ρ_1 with relatively large ρ_2).

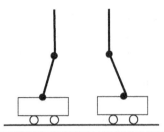

Figure 5.27 Balancing pattern 3 of the double-link inverted pendulum (small ρ_1 with small ρ_2).

ρ_1, the displacement of the car is large. As a result, the system looks like the first pendulum that swings, apparently to keep balance, while the second remains erect.

5.3.4.3 Balancing Patterns of the Triple-Link Inverted Pendulum

There are three association degrees (ρ_1, ρ_2, ρ_3) in the triple-link inverted pendulum system, and each has the value of "large" and "small," so there will be eight combinations altogether:

1. Large ρ_1, large ρ_2, large ρ_3;
2. Small ρ_1, large ρ_2, large ρ_3;
3. Large ρ_1, small ρ_2, large ρ_3;
4. Small ρ_1, small ρ_2, large ρ_3;
5. Large ρ_1, large ρ_2, small ρ_3;
6. Small ρ_1, large ρ_2, small ρ_3;
7. Large ρ_1, small ρ_2, small ρ_3;
8. Small ρ_1, small ρ_2, small ρ_3.

According to the analysis of the double-link inverted pendulum, the association degree of the upper pendulums must be greater than that of the lower pendulums so as to keep the balance. This rule is also applicable to the triple-link inverted pendulum system, so the possible balancing patterns are 1, 2, 4, and 8.

1. *Balancing pattern 1*: Large ρ_1, large ρ_2, and large ρ_3, as shown in Figure 5.28. In this case, the three association degrees are all large, and the displacement of the car is small. Hence, the system looks like an erect single-link inverted pendulum.
2. *Balancing pattern 2*: Small ρ_1, large ρ_2, and large ρ_3, as shown in Figure 5.29. In this case, the association degree between the car and the first pendulum is small, while those between the pendulums are large. The displacement of the car is large. The system looks like a swinging single-link pendulum.

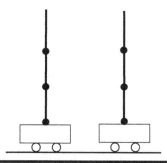

Figure 5.28 Balancing pattern 1 of a triple-link inverted pendulum (large ρ_1, large ρ_2, large ρ_3).

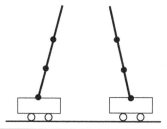

Figure 5.29 Balancing pattern 2 of a triple-link inverted pendulum (small ρ_1, large ρ_2, large ρ_3).

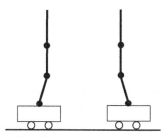

Figure 5.30 Balancing pattern 3 of a triple-link inverted pendulum (small ρ_1, small ρ_2, large ρ_3).

3. *Balancing pattern 3*: Small ρ_1, small ρ_2, and large ρ_3, as shown in Figure 5.30. In this case, the two lower association degrees are small, while the upper one is large. The displacement of the car is large. The upper two pendulums remain erect throughout the process, while the first pendulum apparently swings.

4. *Balancing pattern 4*: Small ρ_1, small ρ_2, and small ρ_3, as shown in Figure 5.31. In this case, all the connections in the joints are soft, so the amplitude of the system is relatively large and the swinging frequency is relatively slow.

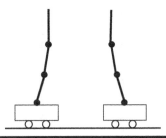

Figure 5.31 Balancing pattern 4 of a triple-link inverted pendulum (small ρ_1, small ρ_2, small ρ_3).

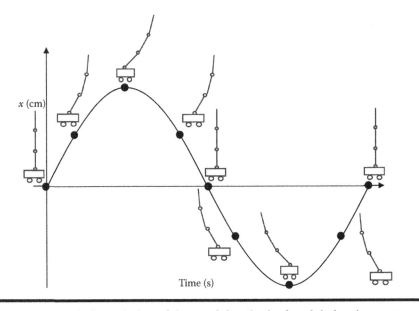

Figure 5.32 Periodic variation of the pendulum in the fourth balancing pattern.

Figure 5.32 demonstrates the periodic variation of a triple-link inverted pendulum when it is in the fourth balancing pattern. The attitudes are different at different times. The angles between the pendulums vary with time; however, in all circumstances the upper association degree is greater than the lower one so as to keep balance.

Figure 5.33 shows the relation between the displacement of the car and the time for the four classical balancing patterns, and the various amplitudes and periods in different patterns.

To balance the triple-link inverted pendulum, the controlling force F on the car is

$$F = F_1 + F_2 + F_3$$

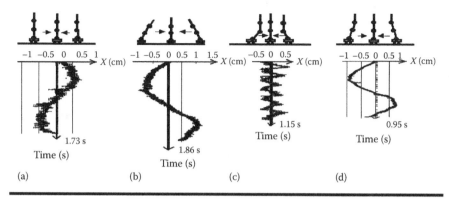

Figure 5.33 **Relation curves between car displacement and time for a triple-link inverted pendulum in different balancing patterns: (a) pattern 1, (b) pattern 2, (c) pattern 3, and (d) pattern 4.**

where F_1, F_2, and F_3 are the controlling forces for the first, second, and third pendulums, controlled by the rule sets $RS(\theta_1)$ and $RS(\dot{\theta}_1)$, $RS(\theta_2)$ and $RS(\dot{\theta}_2)$, $RS(\theta_3)$ and $RS(\dot{\theta}_3)$, respectively. The association degrees ρ_1, ρ_2, and ρ_3 can be changed by modifying the cloud parameters in $RS(\theta_1)$, $RS(\dot{\theta}_1)$, $RS(\theta_2)$, $RS(\dot{\theta}_2)$, $RS(\theta_3)$, and $RS(\dot{\theta}_3)$. By this means the four classical balancing patterns can be constructed.

By using the cloud rule generator and reasoning mechanism, the single-link, double-link, and triple-link have been successfully stabilized by a single motor, which can also maintain balance of the rail for a long time. If the computer manually intervenes and changes the digital characteristics of different cloud rules, the balancing pattern can be dynamically switched, as exhibited in the 14th IFAC Congress. Up to now, there have been no reports about the balancing pattern in inverted pendulum control.

Figure 5.34 records the dynamically switching curve in process pattern 1 → pattern 2 → pattern 3 → pattern 4 through modifying the cloud parameters by computer.

It is of great significance to study the realization, analysis, and induction of the balancing patterns of the inverted pendulums. By modifying the cloud parameters, the motor voltage can be adjusted so as to control the attitude of the system, which provides a possible solution for the operational attitude setting of the controlled object in real applications. Meanwhile, the research of balancing patterns has not only deepened the development of the cloud, but also broadened the cloud application.

To sum up, reasoning and control based on qualitative knowledge are important tools for intelligent control, and a significant part of the research on AI with uncertainty. By introducing the cloud-based uncertainty reasoning and control mechanism, we have solved the stabilization problem in the triple-inverted pendulum system, which is a classical issue in intelligent control. The cloud-based mechanism

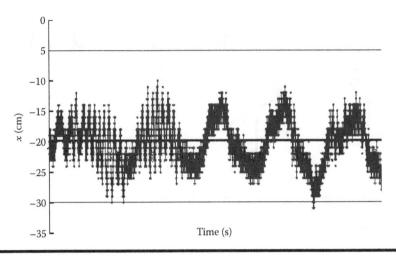

Figure 5.34 Dynamically switching curve between different balancing patterns of the triple-link inverted pendulum.

is explicit and direct, exclusive of complex reasoning and computation, so it can simulate the uncertainty in human thought activities, and is of applicable value.

5.4 Uncertainty Control in Intelligent Driving

With the development of automotive electronics, mobile Internet, cloud computing, and Internet of vehicles, it is inevitably introducing entertainment and information into cars; automotive electronics has changed from the mechanical system-based improvement of car performance, to assisting driving so as to enhance the quality of traveling life. A large number of intelligent driving technologies have improved both people's driving experience and driving safety, such as intelligent receiving radar alarms that help drivers with distance judgment and parking; anti-blink technology that can monitor the driver and then alert them if they are dozing, hence reducing driving accidents caused by fatigue; anti-rear-end technology that starts to ring when the distance is less than safe and forces a wider distance between two cars, hence reducing road traffic accidents.

At the same time, automobiles are facing unprecedented creativity, aiming to become wheeled robots to maximize the interior space without a driver. In 2010, Google announced a project to develop a self-driving car, which has successfully run more than 140,000 miles on roads. In July 2010, the intelligent car, made by Palma, traveled more than 13,000 km from Parma, Italy to Shanghai, China. This magical trip lasted 3 months. In China, the "Future Challenge" match for intelligent vehicles has been successfully held for four versions. Currently, almost every traditional vehicle manufacturer is developing their own smart cars, which have become the focus of intelligent transportation. Intelligent driving will fundamentally change the traditional driving style, free human beings from low-level, tedious, fatiguing and long driving activities, and provide safe and convenient transportation.

In addition, intelligent war-oriented vehicles, when used in high-risk environments, such as battlefields, will effectively extend the range of human activities, reduce human casualties, and showcase more extensive functions. Intelligent driving techniques can also be applied to autonomous unmanned air, surface and submarine vehicles.

5.4.1 Intelligent Driving of Automobiles

The development and research of intelligent vehicles is very complex system engineering based on mechanical and electrical transformation. According to the order of information processing, research and development for intelligent vehicles can be divided into three parts, namely, environmental perception, intelligent decision-making, and automatic control, as shown in Figure 5.35. Among them, environmental perception includes the use of a variety of cameras, laser radars, millimeter-wave radars, infrared radars, and so on, to collect, analyze, and process information from the surrounding environment. We propose the concept of Right of Way and a method for integrating the results from each sensor on the radar map to calculate the right of way to be used by vehicles. Intelligent decision considers the right of way held by the vehicle and its change rate, then uses the rule base to its speed and direction changes in the next moment. The controller enables the vehicle to smoothly reach the desired speed and direction through a suitable method.

5.4.1.1 Integration of the Right of Way Radar Map

Vehicle control is essentially the control of its speed and direction, which must meet the requirements of safe driving. For example, if the distance between a car and the vehicle in front is quite small, and its speed is greater than that of the front vehicle, then the car should slow down. This is the, so-called, right of way concept, the right of the

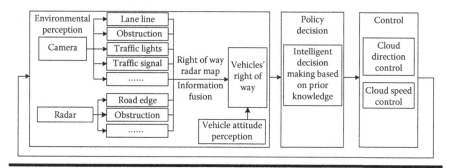

Figure 5.35 Framework for information processing of intelligent driving.

car to occupy the front space at the pace it is moving, which can be divided into the measured and calculated right of way. The measured right of way, as measured actually by the sensors, is the travel space owned by the vehicle; the calculated right of way is calculated using rights of way functions based on the measured right of way and taking into account the vehicle size, quality, speed, surrounding traffic, and so on. The measured right of way provides an important basis for intelligent decision-making and automatic control of unmanned vehicles. The calculated right of way can be used not only to inform traffic jams and characterize their causes in intelligent transportation, but also be used for personalized billing based on different driving behaviors. To put it simply, the calculated right of way is a mobile sector that can be represented by distances and angles:

$$RW(r, \theta) = f(l, v, \xi, r')$$

where

　　r is the length of the right of way
　　θ is the angle calculation of the right of way
　　l is the size of the vehicle
　　v is the current vehicle speed
　　ξ is the surrounding traffic
　　r' is the calculated right of way

This is a nonlinear function of the vehicle speed and its calculated right of way on the road. For an unmanned car, we measure the calculated right of way for intelligent decision making and automatic control. If there is no other instruction, the right of way hereunder refers to the calculated right of way.

　　Essentially, self-driving is the intellectual process of an automobile's constant detection of, request for, and response to right of way; multi-vehicle interaction is the coordination process of a group of automobiles competing for, waiving, and occupying a right of way. Detection of the right of way is both the basis of safe driving and the prerequisite for completing intelligent decision making and automatic control. To accurately detect the right of way in real-time, intelligent vehicles are usually equipped with camera, laser radar, millimeter-wave radar, infrared radar, and other sensors. Effectively integrating a variety of sensors for real-time data is an important and difficult point of intelligent driving technology. For this reason, we built the basic framework of a radar map for right of way, as shown in Figure 5.36. This map covers a circular area with a radius of approximately 200 m. According to human driving experience, the area near the vehicle needs more careful attention the area further away. Based on such granularity, we divided the coverage into 141 rings, with the radial length gradually increasing from inside to outside. The radial length of the outermost ring is about 3.39 m, which reflects coarse-grained attention; the radial length of the inside ring is about 5 cm, which reflects fine-grained attention. The angular resolution of the radar map for right of way is 1°, which divides every ring into 360 different grids.

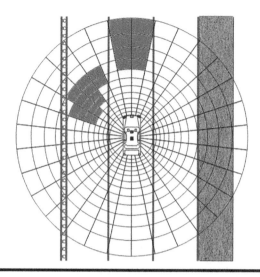

Figure 5.36   Radar map for right of way.

The environmental information perceived by different sensors is mapped on the radar map for the right of way according to their positions through parameter setting and calibration of the sensor. To ensure safe driving, a radar map grid is marked 1 if it is detected by the sensor as occupied by an obstacle, which means the grid cannot be passed through; if no occupation has been detected by all sensors, it is marked 0, which means that the grid can be passed through. The radar map grids for right of way cover the circular area around the vehicle with a radius of 200 m. By using the radar map for right of way, we can quickly and accurately complete multi-sensor data integration for intelligent driving.

To further facilitate judgment, the radar map for right of way can be divided into eight attention areas, as shown in Figure 5.37. Every attention area is a fan-shape. The right of way of each area can be described using radius l and aperture angle α. $l_i(t)$ describes the right of way of the vehicle with radius l in range i at the t moment. $v_i(t) = \mathrm{d}l_i(t)/\mathrm{d}t$, which $v_i(t)$ describes the speed at time t in the range i. $v_i(t)$ is the relative velocity between car and obstacle, which is an important basis for decision making. If we take $l_i(t)$ and $v_i(t)$ as input, in line with human driving experience, we can build a smart driving decisions rule base, and make vehicle-related decisions accordingly.

Figure 5.38 is an example of the integration in a radar map for right of way. Figure 5.38a shows the vehicle's current status, the white car in the middle of the radar map is the subject vehicle, the gray part on the right is the greenbelt, the road bar is on the left side of the vehicle. Through the sensors, the current radar map for right of way can be represented in Figure 5.38b. Figure 5.38c and e represents the front, left front, and back of the right of way.

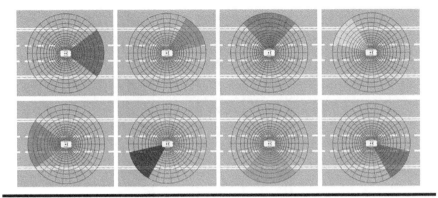

Figure 5.37   Eight areas divided on the radar map for right of way.

5.4.1.2 Cloud Control Strategy of Intelligent Vehicle

Intelligent control of vehicle driving is mainly the control of vehicle speed and angle when it is following the vehicle in front, changing direction, and passing an intersection. In most cases, an intelligent vehicle is in the state of following the one in front. We can take this case as an example and use cloud reasoning and cloud control to intelligently control the vehicle speed and angle.

While following the car in front, the intelligent vehicle will keep adjusting its speed based on information of the car in front, pedestrians, and other obstacles. It will increase driving efficiency and avoid collision, which is also the purpose of intelligent control of vehicle speed. The angle control aims to keep the vehicle in the middle of its lane as far as possible. Just like the principle of the pendulum, by continuously adjusting the steering wheel, the vehicle tries its utmost to make the distance from its center to both the left and right lanes equal, and its driving direction the same as that of the lane in the driving process.

The speed and angle control of the intelligent vehicle is a typical single-input single-output controller. Let us use the angle control as an example in Figure 5.39, the solid line is the actual lane markings, the thick dashed line is the central axis line calculated by the two lane markings. An intelligent vehicle can control the direction based on the distance d between the center of vehicle and central axis line, and angle θ between the driving direction of vehicle and central axis line.

The input of the cloud controller is the distance d between the center of vehicle and central axis line, and angle θ between the driving direction of vehicle and central axis line. The output is the vehicle steering wheel angle δ. Through a brief summary of the human driving experience, we can draw the following qualitative conclusions:

1. If the vehicle does not deviate from the center of the lane, and the vehicle is running in the same direction as that of lane axis, then let the steering wheel back to zero, so that the vehicle runs straight on. That is, if d and θ are close to 0, the δ should be 0°, so that the vehicle stays traveling straight ahead.

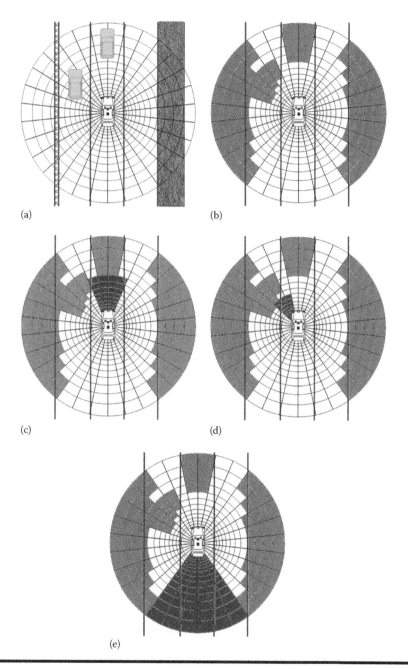

Figure 5.38 Example of integration in radar map for Right of the Way: (a) vehicle's current status, (b) the current radar map for Right of Way, (c) the front of the Right of the Way, (d) the left front of the Right of the Way, (e) the back of the Right of the Way.

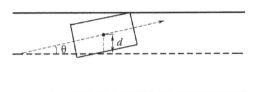

Figure 5.39 Distance *d* between vehicle center and central axis, and angle θ between driving direction and central axis.

2. If the vehicle deviates from the lane center to the right, then turn the steering wheel left, making the vehicle go back to the center of the lane; the more obviously the vehicle deviates, the bigger the angle of the steering wheel should be. If *d* is greater than 0, then δ is less than 0°, and the absolute value of δ is positively related to the absolute value of *d*.

3. If the vehicle deviates from the lane to the left, then turn the steering wheel to the right, making the vehicle turn back to the center of the lane; the more obviously the vehicle deviates, the bigger the angle of the steering wheel should be. If *d* is less than 0, then δ is greater than 0° and the absolute value of δ is positively related to the absolute value of *d*.

4. If the angle between the vehicle running direction and lane axis is greater than 0°, that is, the vehicle is turning right, then turn the steering wheel left, taking the vehicle back to the center of the lane; the more obviously the vehicle deviates, the bigger the angle of the steering wheel should be. If θ is greater than 0°, then δ is less than 0° and the absolute value of δ is positively related to the absolute value of θ.

5. If the angle between the vehicle running direction and lane axis is less than 0°, that is, the vehicle is turning left, then turn the steering wheel right, taking the vehicle back to the center of the lane; the more obviously the vehicle deviates, the bigger the angle of the steering wheel should be. If θ is less than 0°, then δ is greater than 0° and the absolute value of δ is positively related to the absolute value of θ.

Next, we will describe the language value set of input and output language variables, define the scope of each language value, and establish control rules based on the five qualitative rules.

Five qualitative concepts can be used to describe three variables *d*, θ, and δ, namely, {positive large, positive small, close to zero, negative small, negatively large}. Thus, the input and output variables both define the five qualitative concepts to construct the corresponding cloud rule generators.

For vehicles intelligently driving after the cars in front, the direction control rules are listed in Table 5.4. The parameter settings of their qualitative concepts are listed in Table 5.5.

Table 5.4   Control Rules for Intelligent Vehicles

	Rule set RS(d) for distance d from the vehicle center to axis of the lane			
	Distance d to Axis			*Steering Wheel Angle δ*
	Positively large			Negatively large
	Positively small			Negatively small
If	Zero	Then		Zero
	Negatively small			Positively small
	Negatively large			Positively large
	Rule set RS(θ) for the angle θ between the vehicle running direction and axis of lane			
	Axis Angle θ			*Steering Wheel Angle δ*
	Positively large			Negatively large
	Positively small			Negatively small
If	Zero	Then		Zero
	Negatively small			Positively small
	Negatively large			Positively large

Table 5.5   Parameter Settings of Control Rules

	Positively Large	*Positively Small*	*Zero*	*Negatively Small*	*Negatively Large*
RS(d)	(1.5, 0.2, 0.004)	(0.5, 0.15, 0.003)	(0, 0.08, 0.001)	(−0.5, 0.15, 0.003)	(−1.5, 0.2, 0.004)
RS(θ)	(10, 1.2, 0.02)	(5, 1, 0.02)	(0, 1, 0.01)	(−5, 1, 0.02)	(−10, 1.2, 0.02)
δ	(20, 3, 0.05)	(10, 2, 0.02)	(0, 2, 0.008)	(−10, 2, 0.02)	(20, 3, 0.05)

Figures 5.40 through 5.43 show records from intelligent vehicles in a highway driving experiment. Figures 5.40 and 5.41 give the variation curves of the angle between the intelligent vehicle and the lane line, and the distance from the central axis line. As shown in the figure, angle θ keeps stable at −1° to 0.5°, while the distance d stabilizes at 0.1 to 0.7 m, fluctuating slowly within the range of 0.6 m (lane width 3.75 m). Data in this section indicates that the vehicle keeps a good status

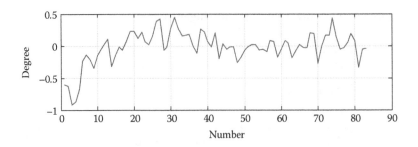

Figure 5.40   Curve of angle between intelligent vehicle and lane.

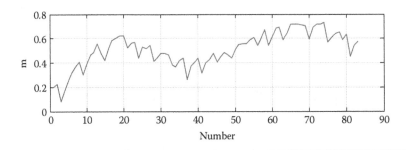

Figure 5.41   Curve of distance from intelligent vehicle to lane axis.

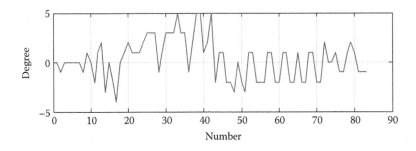

Figure 5.42   Curve of steering wheel angle.

in the lane. Figure 5.42 shows the angle curve of the steering wheel, in which the negative value is for the right direction, and the positive value for the left. The angle of the steering wheel stabilizes within the ranges between –5° to 5°. Figure 5.43 shows the vehicle speed curve.

The intelligent vehicle controller, if it navigates by the lane, can ensure that the horizontal error is less than 20 cm when the vehicle is driving in the virtual center.

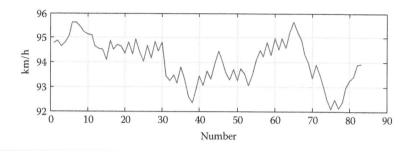

Figure 5.43   Speed curve.

Currently, intelligent driving from Beijing to Tianjin on the intercity high-speed road has been successfully achieved. Under normal weather conditions, it can autonomously follow or overtake the cars in front with a maximum speed of up to 100 km/h.

5.4.2 Driving Behavior Simulation Based on Intelligent Automobiles

With humans, different drivers have different driving behaviors, which can be divided into the three types of new driver, normal driver, and fast speed driver. Under the same road conditions, a new driver is generally more cautious and slow, and neither snatches the right of way from other vehicles, nor overtakes the car in front; while a fast speed driver has a strong desire to snatch the right of way, and will constantly to pursue higher speeds. Even the same type of drivers will show different driving behaviors because of their different personality, level of fatigue, mental state, and so on. This uncertainty of driving behaviors is mainly represented in the uncertainty exhibited in drivers' competition for, abandonment, and possession of the right of way. Even for the same style of the driving, different drivers will show great differences in their behavioral characteristics.

To reproduce traffic flow in the actual environment, a cloud model can be introduced to simulate human driving behavior in intelligent vehicles. On one hand, we can use a backward cloud algorithm to abstract parameters of driving behavioral characteristics from the information on changes in the right of way; on the other hand, we can get different driving behaviors by transforming the characteristic parameters into specific information on right of way with a forward cloud algorithm.

For example, why is the fast speed driver able to keep seizing the right of way from other vehicles, while the new driver is constantly forced to abandon the right of way he had possessed? Because of a specific speed, the fast speed driver dares to hold and release the right of way, but beginners have to possess more right of way. That is, we usually think a fast speed driver has a "bold, careful and fast response." For example, a fast speed driver only needs to occupy a distance 20 m ahead of him to run his car at the speed of 80 km/h; by contrast, the beginners need to occupy a distance of at least 60 m from the car in front to increase the speed to 80 km/h.

To simulate this phenomenon, we carried out an experiment on a miniature intelligent vehicle platform, which is a miniature traffic environment and intelligent vehicle in a ratio of 1:10 with the real environment. They offer a semi-physical research platform for the development of real intelligent vehicles. In the development and testing process of a real intelligent vehicle, dangers can be caused by factors such as un-robust algorithm. So using the miniature platform for algorithm development and debugging can significantly reduce the occurrence of dangerous situations. In addition, research into intelligent transportation requires a true representation of the traffic flow in a real environment. The traffic flow simulated by software is somewhat different from the real situation, but using real vehicles costs much more. Therefore, a miniature platform is a low cost, realistic simulation of traffic flow, which can demonstrate some practical problems, such as traffic intersection reconstruction, vehicle running strategies, and provide a good experimental platform for researching intelligent transportation.

By collecting speed samples from a fast speed drivers, normal drivers and beginners when their cars are 25 m away from the car in front, and using backward cloud generator, we can generate a corresponding speed cloud model (Figure 5.44).

By using the same method, the speed models can be generated when their cars are at different distances. The speed and angle model for changing direction can also be generated. Then we can use the forward cloud generator to generate examples of

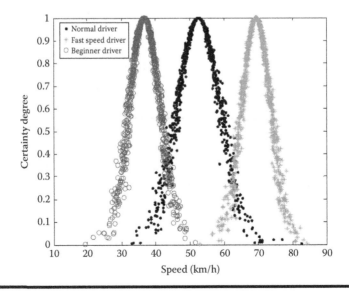

Figure 5.44   Velocity models for drivers of three styles when 25 m from the car in front.

driving behavior and simulate a typical driving behavior, such as following the car in front, lane changing, overtaking, and driving on road intersections, parking, and other acts. Characteristic parameters are different and uncertain in each generation so as to enable the intelligent vehicle to simulate the uncertainty of human driving behavior.

As regards each individual driver, we can collect and accumulate data for each one, and let the intelligent vehicle keep learning the relationship between driver speed and right of way, lane changing tendencies, and other statistical features; then we use them as a cloud drops group for driving behaviors and use the backward cloud generator to obtain digital characteristics of his driving behavior.

This simulation of driving behavior is of practical significance. For example, based on simulated driving behavior, we can study the relationship between right of way and traffic congestion, and find a more realistic solution for dealing with various transportation problems. The core of traffic congestion is the contradiction between supply of and need for right of way. Different driving behaviors use their right of way differently. Due to the excessive pursuit of high speed, the fast speed driver always changes lanes and seizes the right of way from others, resulting in overall lower traffic speed; beginners driving at the same speed need more distance, which not only causes unnecessary waste of road resources, but also reduces the overall driving speed. Because different vehicles contribute differently to road congestion, we can simulate all types of driving behaviors, even an accident scene, to restore true traffic flow. Through an analysis of the occupation of road resources by different drivers and their impact on congestion, we can consider different ways to calculate and collect a congestion charge to ease urban traffic congestion, for instance, we can request a car owner to pay a congestion charge based on the total right of way he uses.

A cloud model can also enable a wheeled robot's self-learning ability. Based on the data collected by human driving behaviors, the expectation, entropy, and hyper-entropy of a specific driver can be obtained by using the backward cloud model. The digital characteristics reflect the basic certainty in uncertainty of driving behaviors. In the course of auto-driving, they can use the forward cloud generator to randomly generate each driving behavior, and their error differences were not statistically significant. As a result, the car can drive like a driver. In this sense, we can also develop a companion robot, racing robots and so on.

References

1. K. Di, The theory and methods of spatial data mining and knowledge discovery, PhD thesis, Wuhan Technical University of Surveying and Mapping, Wuhan, China, 1999.
2. D. Li, Y. Du, G. Yin et al., Commonsense knowledge modeling. In: *16th World Computer Congress 2000*, Beijing, China, 2000.
3. C. S. Roger and B. L. David, Creativity and learning in a case-based explainer, *Artificial Intelligence*, 40(1–3), 353–385, 1989.

4. A. Kandel, A. Martins, and R. Pacheco, Discussion: On the very real distinction between fuzzy and statistical methods, *Technometrics*, 37(3), 276–281, 1995.
5. M. Mizumoto, Fuzzy controls under product-sum-gravity methods and new fuzzy control methods. In: A. Kandel and G. Langholz, eds., *Fuzzy Control Systems*, CRC Press, London, U.K., pp. 276–294, 1993.
6. D. Li, Uncertainty reasoning based on cloud models in controllers, *Computers and Mathematics with Applications*, 35(3), 99–123, 1998.
7. H. Chen, Qualitative quantitative transforming model and its application, Master's degree thesis, PLA University of Science and Technology, Nanjing, China, 1999.
8. N. Zhou, Cloud model and its application in intelligent control, Master's degree thesis, PLA University of Science and Technology, Nanjing, China, 2000.
9. F. Zhang, Y. Fan, D. Li et al., Intelligent control based on the membership cloud generators, *Acta Aeronautica ET Astronautica Sinica*, 20(1), 89–92, 1999.
10. H. Chen, D. Li, C. Shen et al., A clouds model applied to controlling inverted pendulum, *Journal of Computer Research and Development*, 36(10), 1180–1187, 1999.
11. F. Zhang, Y. Fan, C. Shen et al., Intelligent control inverted pendulum with cloud model, *Control Theory and Applications*, 17(4), 519–524, 2000.
12. D. Li, H. Chen, J. Fan et al., A novel qualitative control method to inverted pendulum systems. In: *14th International Federation of Automatic Control World Congress*, Beijing, China, 1999.
13. D. Li, The cloud method and balancing patterns of triple link inverted pendulum systems, *The Chinese Engineering Science*, 1(2), 41–46, 1999.

Chapter 6

Cognitive Physics for Swarm Intelligence

Artificial intelligence has been developing for 60 years. People first focused on individual intelligent behavior, but they are now paying more attention to swarm intelligence, which is difficult for an individual to possess. This chapter looks at swarm intelligence through cognitive physics.

6.1 Interaction: The Important Cause of Swarm Intelligence

Complexity science has developed research methodologies with unique features, and emergence is one of the most important concepts. It is recognized that system characteristics may be derived from different mechanisms. Sometimes, system elements are independent from each other or have a very weak interaction, so they acquire the overall characteristics of a system with the mechanics of linear superposition. At other times the system elements have a strong interaction, and the system's characteristics are determined by the structure created by that interaction. Each element is affected by the surrounding ones, and a weak affect may be amplified quickly, which leads to a great change, so that the system behaves at a higher level and with more coordination and order, which is known as emergence. The emergence of new features can be seen as a kind of swarm intelligence.

6.1.1 Swarm Intelligence

The earliest concept of swarm intelligence comes from the observation of insect populations in nature. In the biological field, swarm intelligence refers to the behavioral characteristics of macroscopic intelligence as exhibited by collaborative, gregarious creatures. Study on a large number of animal groups—including ants, bees, spiders, fish, and birds—has spawned various intelligent computing models. Typical examples are the ant colony algorithm proposed by Dorigo [1] and the particle swarm optimization proposed by Kennedy and Eberhart [2]. At the core of each model is a group composed of many simple individuals that present an overall capacity to complete a complex task through simple mutual cooperation. For example, in Figure 6.1 the ant colony relies on group behavior to find an optimized path. In these groups of creatures, the individual's own structure is relatively simple, but their group behavior presents a complex performance. That performance is not the result of the nature of a single individual, neither can it be easily predicted and deduced by an individual's simple behavior.

In fact, swarm intelligence can also be seen in human society, and is an important part of today's social computing. There is an old saying in China: "Three fools are the equal of one wise man." This is actually the original version of swarm intelligence in human society. Nowadays, people are more and more interested in a variety of swarm intelligence emerging from "small minorities," which relies on Internet-based interaction, and they are conducting research on numerous related social phenomena, such as "crowd-sourcing," "Wikinomics," "human flesh searches," and so on. A group of people who are geographically dispersed but who share a common interest can solve a large number of difficult problems that cannot be solved by computers or individuals through communication via the Internet, and in an amazingly effective way. Swarm intelligence formed this way can even offer huge economic or social value. For example, in Wikipedia individuals in a "small minority" interact

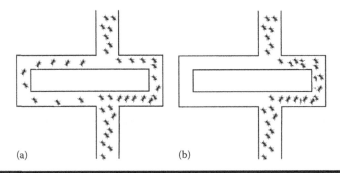

Figure 6.1 Ant colony finds an optimized path: (a) ant colony initial path, (b) ant colony final path.

with each other, collectively create Wikipedia entries, and establish the encyclopedic knowledge; in social networks individuals can create labels or recommendations, forwarding or sounding, which exchanges ideas and makes discussions go with a swing. Groups involved in the sorting of digital resources, quickly complete tasks for identifying, classifying, or clustering pictures, videos, and other digital resources, in a variety of specific situations and themes.

Psychologists and sociologists have long paid attention to the study of group behaviors and the swarm intelligence they present. They study the psychological reactions of human individuals in a variety of environments and any overall characteristics brought out through experiment. However, these experimental results cannot describe the characteristics shown in real-life human interactive behavior, and thus cannot establish a quantitative theory on human behavior. In the context of cloud computing and big data, Internet-based social networks provide a good carrier for the study of human interactive behavior characteristics and the overall swarm intelligence.

A variety of online social networks established via e-mail communications, forum panel discussions, photo sharing (e.g., Flickr), video sharing (e.g., YouTube), microblogging exchanges (e.g., Sina Weibo), virtual social activities (e.g., Facebook, micro-channel, all net) are often found to be "small world" or "scale-free." Their formation mechanism draws the attention of researchers from many disciplines, and the interaction between individuals is one of the fundamental factors. Interaction is the basic form and the formation base of social networks. Interaction between individuals forms a community, and that between communities constitutes a social network. The individual diversity of a social network makes its nodes different from the homogeneous nodes assumed by the classical network model, which is bound to bring diversity to groups. Because individuals may belong to multiple social groups focusing on different aspects in the network, the boundaries between community groups are not clear, and there is a lot of overlap between groups. The freedom and anonymity of individual participation in social networks, the large scale of communities and different individual qualities flood social networks with an enormous amount of invalid information and useless individuals, thus reducing the significant common nature of community groups. In social networks, multiple dominant or recessive interaction forms— such as frequent interaction, incremental interaction, active interaction, extensive interaction, diverse interaction, lasting interaction, and so on—reflect a variety of group behaviors and form different social groups with small and big sizes, coarse and refined themes, and high and low thresholds, while the social group itself can be divided endlessly and the community evolves through growth and decline. Interactions between individuals result in positive or negative interactive influences, which lead to the evolution of communities and networks. The interactive diversity of individuals and groups creates unprecedented difficulties in modeling community evolution, in uncovering community groups and behavioral analysis.

Therefore, quantitative methods, with interaction at the core of the analysis of group behaviors, are the key to the study of swarm intelligence.

In the Internet environment featuring public interaction, the interaction between individuals, and between individuals and the environment, prompts other individuals to make various changes and triggers the emergence of new phenomena at a macro level. The transition phenomenon of macro systematic forms, due to the large number of micro-interactions, is the "emergence" in complexity science.

6.1.2 Emergence as a Form to Represent Group Behavior

The concept of emergence originates from Lewes's philosophical research on life and thinking in the nineteenth century, and is stated in the opening sentence of the chapter on "emergence characteristics" in the *Stanford Encyclopedia of Philosophy*: "Emergence is a notorious philosophical term of art" [3]. The concept of emergence is very complex. The long-forgotten concept was not reawakened until research on complexity science began at the end of the twentieth century. Today, it is no longer mysterious to regard a phenomenon as an emergence. In a group, the overall swarm intelligence shown through the interaction between individuals is a typical emergence process.

The emergence phenomenon covers extensive fields, ranging from physics, chemistry, and thermodynamics to anthropology, sociology, economics, and so on, all of which describe the emergence phenomenon, as shown in Figure 6.2. In nature this includes the group chirp of crickets, frogs, and cicadas in summer, and the synchronized flashing of fireflies. Emergence in human society includes the rapid

Figure 6.2 Widespread synchronization.

convergence of applause in a theater audience, and women living together for a length of time sharing the same menstrual cycle due to mutual influence. There are similar biological phenomena, such as the bioelectrical current produced synchronously by cardiac pacemaker cells to enable the heart to beat rhythmically. Most nerve cells arranged in the same order and the same direction discharging brainwaves synchronously. In a physical community, emergence is a mutation from one state to another, often referred to as phase transition. As early as 1665, physicist Christiaan Huygens discovered that two pendulums hanging on the same rope can swing synchronously after a time.

Study of the emergence phenomenon has always been linked to complex systems. They are some new properties and phenomena produced by interactions between individuals within a complex system, which can only be observed at the scale of the overall system . These properties and phenomena are not under the control of a unified coordination, but emerge from the interaction between individuals in a partial environment. Therefore, emergence is always associated with mutual interaction, which is a necessary condition for emergence. Emergence studies the mechanism of "the whole greater than the sum of its parts" and describes changes of a system from low to high level, from part to whole, and from microscopic to macroscopic, which emphasizes bottom-up mutual interaction. It is this interaction that leads to the appearance of swarm intelligence with certain overall features, and with overall macroscopic properties different from the sum of the individuals' microscopic characteristics.

Emergence is a kind of coupling interaction in which the front is associated with the back. Such an interaction is mainly the self-adaptability and environmental adaptability of individuals, and there is no unified overall control. Technically, these interactions and the system produced by the effect are both nonlinear. Behavior of the whole system cannot be achieved by the simple summation of its various components. We cannot truly understand a chess player's strategy by assembling the statistical values of each move in the chess game, nor can we understand the behavior of a whole ant colony through its average activity, or understand the content of Wikipedia by summarizing the content contributed by the editors. In these cases, the entirety is bigger than the sum of its parts.

In most real systems there is an interaction between individuals, and the group behavior shown in such interactions emerges from its individuals' behavior in a nonlinear manner. There is a coupling between individual and collective behaviors with the collective behavior of all individuals forming the behavior of the group. However, group behavior will affect the individuals' behaviors and can change the environment so that individuals change their way of interacting with other individuals, which, in turn, changes the collective behavior. In other words, through the interaction, emergence is a presentation form of a group behavior.

6.2 Application of Cloud Model and Data Field in Swarm Intelligence

A formalized approach is required to understand uncertainty in the behavior of an individual, the interaction and partial functions between individuals, and how this leads to group behavior. We use cognitive physics—cloud model and data field—to explore how, in nature, a large number of individuals end up by displaying a swarm intelligence.

6.2.1 Cloud Model to Represent Discrete Individual Behavior

A large number of group behaviors are often driven by multiple individual behaviors. If there is no interaction between individuals, individual behavior is independent and, on the whole, shows a dispersed state of disorder. This can only be analyzed with statistical methods when the size of the system is large. Overall behavioral statistical results often approximate Gaussian distribution, which is a group behavior different from that exhibited by "emergence." Obviously, if without interaction, individual behavior in a group and overall dispersion can be represented by cloud model.

Cloud model is a transformation model for the uncertainty cognition between qualification and quantification to realize the uncertainty conversion between qualitative concept and quantitative data. The forward cloud model can perform a qualitative to quantitative transformation and produce cloud drops via cloud digital characteristics. The backward cloud model can perform a transformation from quantitative values to qualitative concept and can convert accurate data into a qualitative concept represented by digital characteristics. In many systems, such as a thermodynamic system, there is no mutual interaction between individuals, and individual behavior is random. If the behavior and characteristics of one individual are represented as a quantitative value of the cloud model, that is, the cloud drops, and the behavior and characteristics exhibited as a whole are represented as a qualitative concept of cloud model, where the cloud model serves as a cognitive model for research into the transformation between qualitative concept and quantitative value, then we can study the macroscopic properties exhibited by the system as a whole through individual behavior and characteristics.

If, as a whole, a system seems to be chaotic, the cloud model can study the macroscopic properties exhibited by the system and obtain the valuable laws. According to the central limited theorem, if a random variable value is decided by the sum of a large number of independent causal factors, and the role of each causal factor is relatively and uniformly small, this random variable is subject to Gaussian distribution. In reality, few variables perfectly comply with a certain distribution, and many factors influencing swarm intelligence may be independent from each other. Some factors may be coupled, or even dependent, therefore they often form a less regular distribution and it is an advantage of statistical methods that they uncover relative regularity from random. Statistics can reveal basic certainty in a lot of uncertainties and summarize the overall features from a large number of superficially disordered individuals. For example, different individuals' cognition of one thing is usually weakly coupled or approximates independence, but all individuals can have a similar cognitive style and behavioral standards. The behavior of each individual may produce different random cognitive results, and the overall features shown by a large number of individuals are subject to a specific law. A concept may be obtained with the cloud model and represented by three digital characteristics. The study of the macroscopic characteristics shown by a large number of independent individuals as a whole reflects the public's swarm intelligence.

6.2.2 Data Field to Describe Interactions between Individuals

To study the interaction between individuals, we first need to know the subjectivity of the individual, then to study the preferential attachment properties and local impact of the interaction. Individuals are autonomous intelligent bodies, equipped with a subjective cognitive ability, with their own thinking and behaviors, and exhibiting different cognitions when dealing with the same things. The cognition subjectivity exhibited in interactions makes differences between different subjects inevitable. The cognitive differences, with varying degrees, between one individual and different others may determine that a single individual will show a greater probability of selecting an individual with a similar cognition for an interaction, showing preferential attachment characteristics, which is another important interactive feature. In addition, the interaction between individuals is affected by the distance between them, namely, the closer distance, the greater the influence, and, the farther the distance, the smaller the influence. Typically, no subject can know the status and behavior of all other individuals. Each individual can only obtain information from a small subset composed of such individuals, so, in terms of processing "local information," an individual can only affect limited objects. Therefore, the influence between individuals rapidly declines with distance and a universal phenomenon is the local influence of characteristics. Overall consensus can be achieved through the interaction between individuals, which is not necessarily an individual's cognition at the initial stages or the average of all individuals' cognition, but more a result of interactive emergence.

Human cognition and the thinking process is essentially a simplifying inducting process from data, to information, to knowledge. The data field approach extends the cognitive theory in modern physics for the objective world to the cognition of the subjective world. It introduces the interaction between individuals, describes in a formalized way the original, chaotic, and deformed association between individuals, and establishes a cognitive field. In the cognitive field, interactive individuals in the system can all be seen as interactive objects in the field space. Data field introduces Gaussian potential function to describe the interaction between individuals, which can be used as a method to describe the interactive influence between individuals in the system, where the quality factor reflects the subjectivity of the individual.

The greater the quality, the stronger the reflection of the individual's interactive ability. The greater the distance between the individuals, the smaller the mutual influence. Distance reflects differences between individuals, thus reflecting the preferential attachment characteristics in individual interactions. Interactions between individuals decays fast with distance, and data field describes such local influence characteristics of individual interactions by controlling the attenuation factor of the Gaussian potential function. Therefore, the subjectivity, preferential attachment, and local impact exhibited by individuals in an interaction can be described by the data field.

6.3 Typical Case: "Applause Sounded"

Let us start with one of the phenomena in our daily life.

Spontaneous applause from audiences in a concert hall may sometimes change from chaos to synchronization, which is an emergence phenomenon of a complex system in a human cognitive dimension. It has drawn a lot of people's attention and become a classic case study. Let us envisage a concert hall accommodating an audience of thousands, after the start of a wonderful performance the audience is seated quietly, watching the plot develop, and with the arrival of the climax or at the end of the performance, they burst into thunderous applause to show their appreciation and to encourage the actors. Each member of the audience hears the applause within a certain range and the clapping affects each individual; sometimes the applause is disordered at the beginning and will gradually become rhythmic. It seems that a mysterious force drives the audience to applaud at a consistent rhythm, this is an occurrence of the emergence phenomenon. Although it is difficult for this kind of phenomenon to emerge in everyday life, it does occur occasionally.

6.3.1 Cloud Model to Represent People's Applauding Behavior

The spontaneous synchronous applause in a concert hall is a recognition of the performance. Each member of the audience in the hall is a subject with parameters of age, gender, mood, hobbies, and other aspects; the hall's size, sound, lighting, and other environmental factors can also cause subtle differences. To study problems related to complex systems, a necessary simplification is indispensable, which is also an effective means for revealing the nature of a problem.

6.3.1.1 Simplification and Modeling of Individual Behavior

An individual audience member applauding in a concert hall, after completing his first clap, will continue, at short intervals, to clap for a second, third, fourth and more times until the applause is completed. The continuous applause of an individual can be represented as a continuous square wave signal, as shown in Figure 6.3.

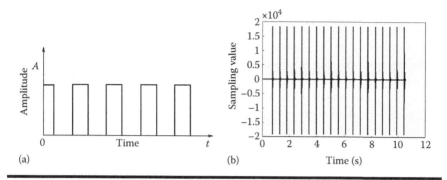

(a)   (b)

Figure 6.3 (a) Square wave signal, and (b) waveform of the continuous applause by an individual.

Signals can usually be represented by cycle, phase, and amplitude. Using this description method, and in order to distinguish, we use clap starting time, clap interval, and applause intensity to represent the applause behavior of an audience [4]. Starting time refers to the time of the first clap from the audience, represented by t_1; clap interval is the time gap between two claps and we use Δt_i to represent his ith of clap interval ($i = 1, 2, \ldots, n$); the applause intensity is used to measure the intensity of the audience's applause, denoted as Q. These three parameters correspond to the initial phase, the cycle, and the amplitude of a periodic signal, respectively, as shown in Figure 6.4.

When an audience watches a performance in a concert hall, and the performance reaches a climax or comes to an end, the audience will applaud quite quickly one after another. Because of physical and psychological differences, the first clap time (starting time) for each audience member is not the same, neither is the initial clap interval for each audience member. So, overall, it sounds chaotic, but through the interaction between the audience members, the applause shifts from chaos to synchronization.

In addition to the three parameters of starting time, clap interval, and applause intensity, which describe each individual audience member's applause, there is

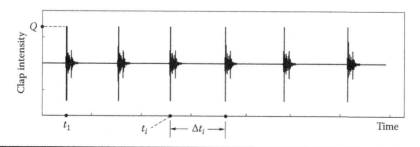

Figure 6.4 Parametric representation of an individual's applause.

another important parameter, the number of claps. Each audience member applauds a different number of times, and they are also influenced by the atmosphere. These four parameters describe the behavior of an individual when clapping. To facilitate the following description, we introduce a new parameter of the clap moment, which is not independent and is decided by the starting time and clap interval of the individual. The clap moment refers to the audience member's starting time and the sum of all clap intervals. For example, the first clap moment of the audience member is his clap starting time t_1, and the second clap moment is $t_1 + \Delta t$, and so on; the ith of clap moment is $t_i = t_1 + \sum_{k=1}^{i-1} \Delta t_k$.

The study of the overall applause in the concert hall focuses on the applauding behavior of each member of the audience, including starting time, clap interval, applause intensity, and clap frequency. Like differences in height and weight, each person's starting time, clap interval, intensity and clap times are not the same. Overall, they all approximately obey pan-Gaussian distribution. The differences in age, gender, mood, and other aspects can be ignored.

6.3.1.2 Simplification and Modeling of the Environment

Concert halls vary in size, structure, lighting, and acoustics, which all impact the emergence of synchronous applause. But because these effects are approximately equal for all the audience, they can be neglected in the context of this study. If our aim is to study small and medium sized concert halls, the influence radius of the applause is a factor that must be considered. Each audience member's applause can affect the applause of others within a limited distance range.

6.3.1.3 Initial Distribution and Presentation of Individual Behavior

Each audience member is independent of the other when they are not applauding, and there is no interaction between them. The original clap from each audience member is a single, tiny, and independent event because at that moment the audience has not been affected by each other's applause. For each audience member, the starting time, clapping interval, applause intensity, clapping times, and the initial value may be considered to obey approximately a pan-Gaussian distribution. Therefore, we can use a cloud model to generate the audience's initial state, and a Gaussian cloud generator to obtain the initial values for starting time, clap interval, clap intensity, and applause times.

In daily life, a normal individual applauds about twice per second. Let us assume the expected normal clap interval to be 500 ms; and when a performance reaches a climax or comes to an end, the audience will react after 1 s, on average, the average value of the starting time of all audiences is about 1000 ms; the average value of applause intensity is approximately 50 dB. We can select the appropriate entropy and hyper entropy, and use a Gaussian cloud generator to obtain the initial values of starting time, clapping interval, clap intensity of each audience member, to reflect

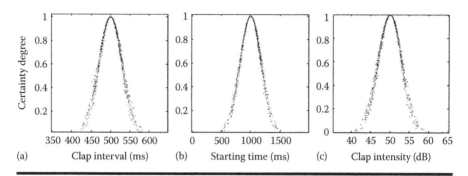

Figure 6.5 The individual's initial state when "audiences applaud": (a) *Ex* = 500, *En* = 30, *He* = 2; **(b)** *Ex* = 1000, *En* = 180, *He* = 10; **and (c)** *Ex* = 50, *En* = 3, *He* = 0.2.

the differences. Figure 6.5 shows cloud drops of 1000 individuals in the concert hall generated by the cloud generator after selecting a different entropy and super entropy for their starting time, clap interval, and clapping intensity. Each cloud drop corresponds to one individual and more cloud drops can naturally be generated.

6.3.2 Data Field to Reflect Mutual Spread of Applause

From the perspective of social psychology, the clapping synchronization formed in the concert hall is caused by a herd mentality. Conformity is a fundamental mechanism of social psychology, shown in people's pursuit of clothing, work options, house purchase, shopping, and so on, as is the mutual influence of applause in the concert hall. When a performance reaches its climax or comes to an end, audiences spontaneously applaud the wonderful performances. The speed and intensity of the clapping are not the same. If an audience member hears that the applause around him is faster than his, he will often speed up his applause frequency. If he hears that the applause around him is more intense than his, he will applaud with more effort. As a result of a constant mutual influence, ultimately, it is possible to make audiences applaud synchronously.

During the clapping process there are differences between individuals in terms of clapping interval and clapping moment, as shown in different frequencies and phases. Due to the influence of herd mentality, the audience will adjust their next clapping interval spontaneously, according to the surrounding applause speed or time, which will in turn affect others. The clap interval of each audience member is adjusted based on their own last clap interval, and according to those by people around them. Their own clap interval and moment will affect the next clap interval of those around him. Eventually everyone's clap interval and clap moment will tend to synchronize, so the emergence phenomenon appears.

Applause intensity decays with distance, which is an important factor in mutual influence between audience members. Each member has a radius of influence in the concert hall, which can only affect others within a certain range, and he is only

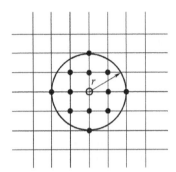

Figure 6.6   Range of audience mutual influence (*r* = 2).

affected from within that range. The mutual influence between him and the audience beyond a certain range is almost zero. Let us assume the influence radius of each individual is the same, but because of the different intensity applause of each, they have a different ability to affect other individuals. Figure 6.6 shows an example with an influence radius of *r*. A person can affect 12 people around him, and the people represented by the hollow nodes are also affected by others around them, represented by solid nodes with *r* as radius. Each audience member is affected by other members around them who, at the same time, influence those around them. There is no overarching control to the whole process, but only a limited environmental interaction between some audience members.

According to this model of influence between individuals, field may be used to make the description. According to the data field theory, the applause of each audience member can be seen as an object whose subjective behaviors contain the three constantly changing parameters of clap moment, clap interval, and clap intensity. The audience's applause spreads through clapping intensity, and the greater the clap intensity, the farther it travels, and the greater the influence it will exert on the surrounding audience. Therefore, clapping intensity can be used as the object's quality, and the physical distance between two audience members as the distance between them. Thus, a uniform and symmetric field will be formed around each object. The potential formed by one object at the position of other object can be calculated by a data field [5].

6.3.3 Computing Model for "Applause Sounded"

According to the data field theory, there are $M \times N$ lines of seats in a concert hall, and audience members form a $M \times N$ matrix. The audience located at $[i, j]$ and $[x, y]$ are denoted by $A_{i,j}$ and $A_{x,y}$ in the $M \times N$ matrix, the corresponding potential of the applause of audience member $A_{i,j}$ spreading to the position of $[x, y]$ is

$$\varphi_{i,j}\left(x, y\right) = Q_{i,j} \times e^{-\frac{(x-i)^2 + (y-j)^2}{\sigma^2}}$$

where

$Q_{i,j}$ is the clap intensity of member $A_{i,j}$

σ is the distance influence factor, and $\sigma = \left(\sqrt{2}/3\right)r$ (for a given σ, the influence area of Gaussian function approximates $\left(3/\sqrt{2}\right)\sigma$), $1 \le x \le M$, $1 \le y \le N$, $1 \le i \le M$, $1 \le j \le N$

Let us define $S_{x,y}$ as the set of all the other members whose distance from $A_{x,y}$ is less than the influence radius r, namely, $S_{x,y}$ is the set of audience members around $A_{x,y}$, then

$$S_{x,y} = \left\{ A_{i,j} \,\middle|\, \sqrt{(x-i)^2 + (y-j)^2} \le r \right\} \quad x \ne i, \; y \ne j$$

the applause of each member in the set $S_{x,y}$ generates an impact on member $A_{x,y}$, whose potential on position $[x, y]$ can be summed up by the following formula:

$$\varphi(x,y) = \sum_{A_{i,j} \in S_{x,y}} \left(Q_{i,j} \times e^{-\frac{(x-i)^2 + (y-j)^2}{\sigma^2}} \right)$$

The next clap interval of audience member $A_{x,y}$ is $\Delta t'_{x,y}$, which is codetermined by his own previous clap interval $\Delta t_{x,y}$ and the most recent clapping behaviors of surrounding members $A_{i,j}$ ($A_{i,j} \in S_{x,y}$). The clapping behaviors of the audience members consist of clapping speed and time. Audience member $A_{x,y}$ adjusts his clapping interval according to the gap between the average clapping interval of the surrounding audience and his, as well as the gap between their average clap moment and his. If the average clapping interval of the surrounding audience is smaller than his, he will reduce his clapping interval, or if it is bigger, he will increase his clapping interval. If the average clapping moment of surrounding audience is earlier than his, he will reduce his clapping interval, or if it is later, he will increase his clapping interval. So the next clapping interval can be formalized as

$$\Delta t'_{x,y} = \Delta t_{x,y} + \frac{\displaystyle\sum_{A_{i,j} \in S_{x,y}} \left\{ \begin{array}{c} \left[c_1(t)\left(\Delta t_{i,j} - \Delta t_{x,y}\right) + c_2(t)\left(t_{i,j} - t_{x,y}\right) \right] \\ \times Q_{i,j} \times e^{-\left(\left((x-i)^2 + (y-j)^2\right)/\sigma^2\right)} \end{array} \right\}}{\displaystyle\sum_{A_{i,j} \in S_{x,y}} Q_{i,j} \times e^{-\left(\left((x-i)^2 + (y-j)^2\right)/\sigma^2\right)}}$$

where $c_1(t)$ and $c_2(t)$ are coupling functions. $c_1(t)$ reflects the influential strength of the clapping interval of the surrounding audience $A_{x,y}$. $c_2(t)$ reflects the influential strength of the clapping moment of those surrounding $A_{x,y}$ on him. $\Delta t_{i,j}$ and

$\Delta t_{x,y}$ are the last clapping interval of members $A_{i,j}$ and $A_{x,y}$, respectively. $\Delta t'_{x,y}$ is the next clapping interval of $A_{x,y}$. $t_{i,j}$ and $t_{x,y}$ are the last clapping moment of $A_{i,j}$ and $A_{x,y}$. The denominator is the potential formed by the applause of all the audience around $A_{x,y}$, which can be normalized.

The above equation shows that each audience member is only affected by other members within the range of influence radius r, that is, each audience member can only use local information around him to adjust himself. There is no overarching control to the whole process, nor is there a pair of invisible "hands" commanding the audience. Each audience member affects the surrounding audience members while equally being affected by them. Since the initial state of each audience member is generated by the cloud model, the model also fully considers the randomness of the individual's initial state.

Obviously, the times of audience clapping has a direct impact on whether synchronization can be formed. If the times of all the audience clapping are less, even if the coupling is very strong, there is no simultaneousness. Only when there are sufficiently more clapping times, will spontaneously synchronized applause be formed, due to the mutual coupling effect. In addition, the audience will not keep applauding all the time. When the atmosphere is no longer hot, the audience will gradually stop applauding. In general, at the climax or the end of a performance, each audience member will have an initial value of clapping times, which is subject to a pan-Gaussian distribution. On this basis, the clapping times of a single audience member is regulated by the atmosphere influence factor, which can be measured by the orderliness degree of the audience's clapping in a concert hall. So the changes in audience clapping times can be expressed as

$$L_{x,y}\left(t+1\right)=L_{x,y}\left(t\right)+c_3\left(t\right)\times L_{x,y}\left(t\right)=L_{x,y}\left(t\right)+K\times\left(\mathrm{std}\left(t-1\right)-\mathrm{std}\left(t\right)\right)\times L_{x,y}\left(t\right)$$

where $L_{x,y}(t+1)$ is the total applauding times of audience member $A_{x,y}$ after being affected by the atmosphere influence factor. $L_{x,y}$ is the applauding times of member $A_{x,y}$ before being affected by the atmosphere influence factor. $c_3(t) = K(\mathrm{std}(t-1) - \mathrm{std}(t))$ is the atmosphere influence factor, which changes over time, and is directly linked to the surrounding applause situation. $\mathrm{std}(t-1)$ is standard deviation of clap moment at the time $t-1$. $\mathrm{std}(t)$ is standard deviation of clap moment at time t. K is the normalization coefficient.

When the calculated atmosphere influence factor is less than 0, the applauding times of all audience member will not increase any more. In fact, it is unnecessary to calculate how many applauding times are added after each clap. When the applauding times of an audience reach the last calculated total clapping times, we can calculate the addition value of the next time. If the addition value is less than 0, the applauding process ends.

In general, the coupling functions $c_1(t)$ and $c_2(t)$ reflect how the interaction between audiences changes over time, and $c_3(t)$ reflects changes in the on-site atmosphere.

In particular, when $c_1(t)$ and $c_2(t)$ are small, or are even 0, it means that almost no interaction occurs between the audience. Thus is general applause formed, which is also known as courteous applause, that usually occurs on ceremonial occasions or as needed by the particular scenario. In studying the emergence process, we have found that a constant coupling function can be used to simplify a problem in many cases when such a coupling function becomes a coupling coefficient.

6.3.4 Experimental Platform

Based on the formalized computing model of "applause sounded," we set up an experimental computing platform using computer programs to simulate the formation of synchronized applause and to further study the mechanism for the emergence of synchronized applause. What should be pinpointed is that applause is just a carrier, and it is easy to load the individual's single applause recorded by digital audio tools into the data stream composed of everyone's clap start time and clap interval in one audience so as to generate the individual's continuous applause, and the clapping sound of many audiences, by synthesizing more individuals' continuous applause. Thus, what is important is to obtain the data stream formed by the clap interval of each audience under the interaction with each other.

To simplify how the data stream in an experimental platform is generated, we assume an audience matrix with $M = N = 8$ (Figure 6.7a), the initial state of their clapping is generated by a cloud model. When applause is heard, the first applauding audience member is found according to everybody's clapping moment and his next clap interval is calculated as the above formula. Then the next first applause of an audience member is found, according to their clapping moment, and the process is repeated until the clapping comes to an end. As shown in Figure 6.7b, the data stream for the clapping interval of 13 audience members in position (4, 5) and the neighboring positions in the 8×8 matrix is withdrawn. All the clapping intervals of the audience members represented by each row from clap start to end of claps are marked with numbers with a unit of ms, the lateral axis of the little rings denotes the actual clapping moment. Let us take $A_{4,5}$ as an example where the first number 249 ms represents the clap start time, the following numbers are his clapping intervals of 345 ms and 363 ms; the number in the little rings is his clapping order within the total 13 clapping audience members, that is, the 1st, 2nd, and 3rd claps of $A_{4,5}$ are, respectively, the 2nd, 14th, and 27th claps of these 13 audience members. The synchronized clapping sounds very clearly when the clapping moments of all audiences are close enough to each other.

According to the above algorithm, it is easy to extend the generation of the data stream to a larger-scale matrix, such as 32×32 or 64×64 audience members. The generation process for data stream files reinforces that there are no invisible "hands" to control the synchronization of the audience applause.

According to the initial state and the audience interaction as regards the "applause sounded," the computing experimental platform for "applause sounded" can be

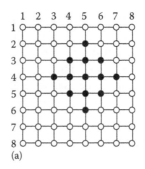

(a)

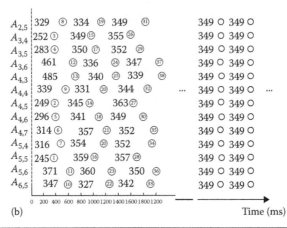

(b)   Time (ms)

Figure 6.7   Generation of data stream file: (a) 8 × 8 audience matrix, and (b) data stream for clap intervals of the 13 audiences.

established, shown in Figure 6.8. The upper left corner is the parameter setting and control button of the platform. The upper right corner is multipeople synthetic applauses. The rows in the lower part are the applause matrix of each audience, such as applauses of 64 (8 × 8), 256 (16 × 16), or 1024 (32 × 32) audience members. The rectangular grid in each row is the waveform of the continuous applause of one audience member. We can select any combined area in the applause matrix by using a mouse, and the synthetic applause effect can be obtained in the upper right corner. The experimental platform records the data stream for clapping intervals of each audience member, which can be used to calculate the corresponding clapping moments. Then the recorded single applause of a single member can be loaded on to these moments in the data stream file to form the file for the continuous applause of a single audience member. Then we use a multimedia mixing algorithm to synthesize files on a single person's continuous applause for all audience members to obtain applause files for many individuals and play them through a digital audio player.

By changing the initial experimental parameters, we can get different types of applause. By selecting the appropriate $c_1(t)$, $c_2(t)$, we can obtain spontaneous

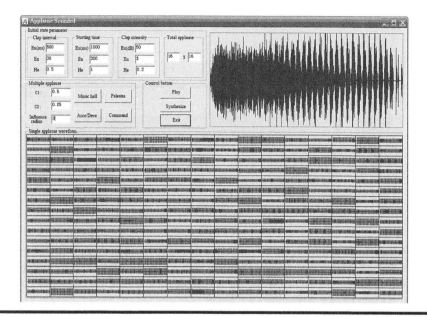

Figure 6.8   Computing experimental platform for applause sounded.

synchronized applause sounds from chaos to orderliness. If we allow the coupling coefficient $c_1(t) = c_2(t) = 0$, at this moment, the calculation formula for interval becomes $\Delta t'_{x,y} = \Delta t_{x,y}$, that is, each audience member applauds autonomously, without being influenced by those surrounding them. There is no interaction between the audience members. Such a simulation is commonly referred to as courteous applause, which often occurs in everyday life. When we change $c_1(t)$, $c_2(t)$ to a small value, it reflects a weak coupling interaction between the audience members and we get messy applause without any orderliness. If the coupling function is enhanced, the synchronization will come but has not yet been realized. When $c_1(t)$ and $c_2(t)$ increase to a certain value, the interaction between the audiences reaches a certain limit and the applause synchronization suddenly appears, thus spontaneously synchronized applauses are formed. Our experimental platform vividly demonstrates the synchronization process, showing applause changing from disorder to orderliness.

Figure 6.9 visualizes the applause of 1024 (32 × 32) audiences. Every node represents an audience member, and the applause of each member is regarded as the vibration of a node. The vertical line represents the vibration amplitude, and the bottom plan represents the matrix of the applauding group. When the amplitude size is positive, the node is deep gray; when the amplitude is negative, the node is light gray.

Considering the randomness, each applause generated by the experiment will not be identical. However, by performing experiments in a concert hall, comparing it with real audience applause, and asking people to differentiate them, we

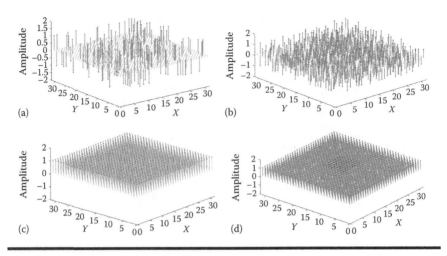

Figure 6.9 Visualized representation of four kinds of applause: (a) courteous applause; (b) interwoven applause; (c) spontaneously synchronized applause; and (d) synchronized applause under guidance.

find that the human ear finds it difficult to identify which is real and which is false. Figure 6.10 shows waveforms of four types of applause, in which the left two pictures are courteous applause, the right two pictures show spontaneously synchronized applause, the top two pictures show the applause recorded on site and the bottom two pictures are applause generated by computing simulation. The waveform of the recorded applause is close to that of the virtual applause generated by computer, indicating that the simulation of the experimental platform is a success.

6.3.5 Diversity Analysis of Emergence

There is a transition from chaos to orderliness during the emergence of applause. Entropy is a classic method to describe the degree of chaos and orderliness of a thermodynamic system. We can use the entropy theory to analyze the degree of chaos in the process of audience clapping.

Let there be m audience members applauding in a concert hall. In the observation window with the time duration of ΔT, we statistically analyze the data stream of clapping moments. At time T, we count the number n of audience members whose clapping moment falls within the observed window ΔT. At time T, the probability of applause is defined as

$$P(A) = \frac{n}{m}$$

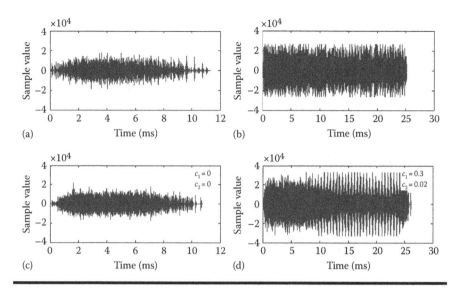

Figure 6.10 Comparison of virtual with real applause: (a) live recorded courteous applause; (b) live recorded spontaneously synchronized applause; (c) computer simulated courteous applause; and (d) computer simulated spontaneously synchronized applause.

The probability of no-applause

$$P(\bar{A}) = \frac{m-n}{m}$$

According to this definition, we know that $P(A) + P(\bar{A}) = 1$. The applause moment time T can be divided into two cases: applause and no-applause. Entropy can be used to analyze the confusion degree of clapping. Within time ΔT, the confusion degree of clapping in a concert hall can be expressed by entropy as [6]

$$H = -\Big[P(A) \times \log_2 P(A) + P(\bar{A}) \times \log_2 P(\bar{A}) \Big]$$

If all audience members clap or do not clap at the same time, then $P(A) = 1$ or $P(A) = 0$. At this moment, $H = 0$ and the applause is synchronous (orderly). However, if half the audience clap, then $P(A) = P(\bar{A}) = 0.5$, $H = 0.7$, the maximum value, and the applause is disorderly. Therefore, the change of H (entropy value) can reflect the chaos and orderliness of an audience's applause. $H = 0$ means most orderly, and $H = 0.7$ means most confusing. Entropy $H = 0$ serves as the standard to judge when applause changes from chaos to synchronization.

We can select a period of typical courteous applause and spontaneously synchronized applause, compare and analyze them with entropy, as shown in Figure 6.11. According to the foregoing discussion on entropy value, we know that when entropy $H = 0$, applause is most orderly and they reach synchronization. Disappearance of the H curve represents the end of the applause process and the disappearance of applause. Comparative analysis of the data stream in Figure 6.11 shows that when it is courteous applause, the behavior of applauding is only limited to the audience members' own states, without coupling between each other. This kind of applause will quickly change to chaos and come to an end after a time. On the contrary, in the case of spontaneous synchronization, due to a wonderful performance, herd mentality makes the coupling mechanism function. The initial chaos begins to gradually weaken, and orderliness emerges. When the entropy curve reaches 0, synchronized applause will be formed, and when the entropy curve ends, the applauding process is over.

Repeatedly playing the different virtual applause types generated by the emergence calculating experimental platform show that clapping interval obviously slows down after synchronization for spontaneously synchronized applause and this slowing down varies for different coupling coefficients. We selected the experimental data on the spontaneously synchronized applause from three typical concert halls to draw the curve of clapping intervals versus time, as shown in Figure 6.12. Each curve in each figure represents the changing trend in an audience's clapping interval; the lateral axis represents time, with unit of s. The ordinate represents a clap interval with units of ms, and the straight line in the figure is the expected clapping interval used by the Gaussian distribution at the initial state.

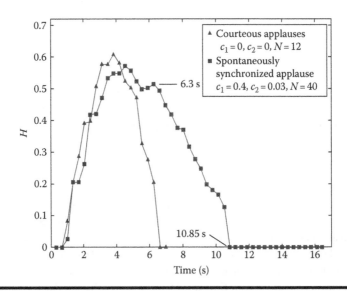

Figure 6.11 Comparison of entropy curves of the two kinds of applause (time window $\Delta T = 400$ ms).

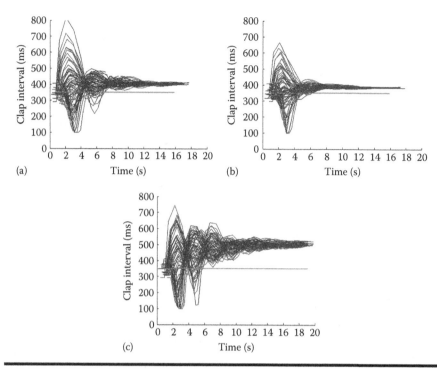

Figure 6.12 Curve of clap interval change: (a) $c_1 = 0.6$, $c_2 = 0.3$; (b) $c_1 = 0.8$, $c_2 = 0.3$; and (c) $c_1 = 0.8$, $c_2 = 0.55$.

From Figure 6.12 we can see that the clapping intervals of all audience members gradually move toward the same after a few times of oscillation, and are significantly larger than the expected value at the initial state. With different coupling coefficients, after synchronization, the slow-down degree of clapping interval is not the same. The emergence is diverse, and there are different kinds of emergence models in reality. So even the clapping interval slows down and the interval may not double, and synchronized clapping may be formed. This is our new discovery from the experiments [7], which is different from the conclusion of Néda in his paper published in *Nature* [8] that only the doubled time interval can form synchronized clapping. Such examples exist in real life, and there are other complex systems that follow this mechanism.

6.3.6 Guided Applause Synchronization

Like many synchronizations of complex systems in reality that require a lot of conditions, in most cases, applause in a concert hall is often difficult to synchronize. Naturally, this would make people ask whether there is a heuristic method that can achieve the synchronization of these complex phenomena and how we can simulate it if it does exist.

If we arrange "shills" among the audience and let them synchronize their applause to influence others without being affected themselves, they will often "guide" the audience to realize a synchronized applause in the concert hall.

If we set the initial values for the shills' applause start time, clapping interval, clap intensity, and clapping times as the expectation value for all audience members, and in the process of the audience clapping, these shills in their "immunity" status will affect the clapping by other audience members, but will not be influenced by them. Ultimately the clapping frequency of all audience members will equal the initial expected value. In this case, by controlling the behavior of the shills, the audience will be guided to quickly reach applause synchronization. The number of shills has a significant impact on the speed of synchronization.

We can select a period of spontaneously synchronized applause without shills, synchronized clapping with 30 shills, and spontaneously synchronized clapping with 60 shills, with the shills randomly scattered among the audiences. The entropy defined in Section 6.3.5 is used to analyze the data stream of the synchronized applause in these three groups, with results shown in Figure 6.13. According to Section 6.3.5, when the entropy is equal to zero, the applause reaches synchronization and when the entropy curve disappears, the applause stops. Thus, it can be found from Figure 6.13 that the synchronization speed is slowest without shills, and the more the number of shills, the faster the synchronization.

Both the number of shills and their arrangement in the audience have significant impacts on the synchronization speed. We can select synchronized applause represented by concentrated shills, randomly arranged shills, and evenly distributed

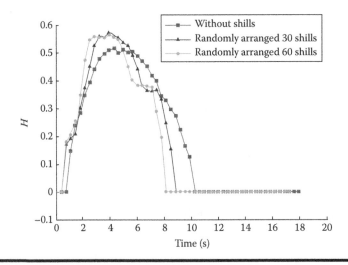

Figure 6.13 Impact of different number of shills on synchronization ($c_1 = 0.4$, $c_2 = 0.05$, $E(\Delta t) = 500$ ms).

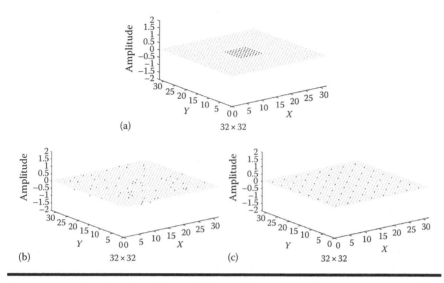

Figure 6.14 Distribution of three different kinds of shills: (a) concentrated shills, (b) randomly distributed shills, and (c) evenly arranged shills.

shills (as shown in Figure 6.14) and comparatively analyze their data streams with entropy, as defined in Section 6.3.5, with results shown in Figure 6.15. The results show that under the centralized distribution of shills, synchronization emerges the latest; synchronization occurs at speeds very close to each other if shills are arranged evenly or distributed randomly, indicating that with the same number of shills, whether or not they are distributed evenly, less impact is exerted on the synchronous speed. It can also be seen from Figure 6.15 that it takes a longer period of time for the one group (community) to spread its message, but if individuals are dispersed among in a larger group, the message can be spread quickly.

Swarm intelligence, as the characteristic or behavior shown by a group as an entirety, is being given more and more attention. If there is no interaction between individuals within the group, the overall features exhibited as swarm intelligence from a statistical point of view can be portrayed by a cloud model with expectation, entropy, and super entropy. If there is interaction between individuals in the group, new properties and phenomena are generated by the system size, and this is also swarm intelligence as the result of interaction between individuals. We used cognitive physical methods to recognize human group behavior and swarm intelligence. We used a cloud model to show individual behavior in a group and the data field to describe the interaction between individuals. We took "applause sounded" as an important carrier to explore the emergence of swarm intelligence.

In the past, people studied biological group behaviors in nature, but now people are increasingly concerned about group behaviors through interaction among human beings in Internet-based social networks, such as open, democratic,

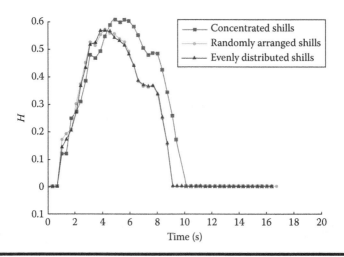

Figure 6.15 **The influence of the difference in the distribution of shills ($c_1 = 0.4$, $c_2 = 0.05$, $E(\Delta t) = 500$ ms).**

bottom-up, self-discipline, self-government, voluntary, and so on. Individuals involved in the interaction are no longer simply a smart organism but show a great thinking ability on a larger scale. The individual's own behavior, and the interaction between individuals, show greater complexity and uncertainty. The influence of individuals can be spread faster and farther through the Internet, and the influence strength may be enlarged continuously, so group behavior is more difficult to predict. These new features make it more difficult for us to understand swarm intelligence, but we also recognize that the intelligence formed by group interactions on the Internet is not centrally controlled, which not only provides new thinking for finding solutions to many complex issues, but also lays the foundation for exploring human interactions on the Internet and the swarm intelligence exhibited as a whole.

Swarm intelligence could expedite a new kind of human calculation model, which is different from past models, including the Turing model. More understanding of swarm intelligence will bring greater progress to human science, so it is worth further exploration.

References

1. E. Bonabeau, M. Dorigo, and G. Theraulaz, *Swarm Intelligence: From Natural to Artificial Systems*, Oxford University Press, New York, 1999.
2. J. Kennedy, R. C. Eberhart, and Y. H. Shi, *Swarm Intelligence*, Morgan Kaufmann Publishers, San Francisco, CA, 2001.

3. E. Zalta, ed., Emergent properties, *Stanford Encyclopedia of Philosophy*, Center for the Study of Language and Information, Stanford University, September 2002.

4. D. Li, K. Liu, Y. Sun et al., Emergent computation: Virtual reality from disordered clapping to ordered clapping, *Science in China*, 51(5), 1–11, 2008.

5. K. Liu, D. Li, Y. Sun et al., Self-organization synchronization research in complex network, Invited Lecture, *National Conference on Complex Networks*, Wuhan, China, 2006.

6. C. E. Shannon, A mathematical theory of communication, *Bell System Technical Journal*, 27(6), 379–423, 1948; (10), 623–656.

7. D. Li, K. Liu, Y. Sun et al., Emerging clapping synchronization from a complex multiagent network with local information via local control, *IEEE Transactions on Circuits and Systems—II: Express Briefs*, 56(6), 504–508, 2009.

8. Z. Néda, E. Ravasz, Y. Brechet et al., The sound of many hands clapping—Tumultuous applause can transform itself into waves of synchronized clapping, *Nature*, 403(6772), 849–850, 2000.

Chapter 7

Great Development of Artificial Intelligence with Uncertainty due to Cloud Computing

7.1 An Insight into the Contributions and Limitations of Fuzzy Set from the Perspective of a Cloud Model

7.1.1 Paradoxical Argument over Fuzzy Logic

It has been more than half a century since fuzzy sets were proposed in 1965 by Professor L.A. Zadeh. Zadeh made pioneering efforts with fuzzy sets to advance cognitive science, and his work is worth recording. Zadeh, with aid of classical mathematics, set up a fuzzy sets theory, proposed thoughts such as language variables, granular computing, soft computing, and computing with words to make an outstanding contribution to the development of cognitive science. However, the fuzzy phenomenon is much more complex than the deterministic phenomenon. Following in-depth study of the fuzziness problem, it was found that fuzzy sets, based on Cantor's set theory, has ineluctable limitations, so it is often subject to criticism.

In 1993, *the 11th National Conference on AI* was held in Washington and Professor Charles Elkan, of the University of California, San Diego, made a presentation entitled

"The paradoxical success of fuzzy logic" [1], which created strong repercussions in the AI community and attracted 15 review papers, including Zadeh's "Responses to Elkan: Why the success of fuzzy logic is not paradoxical" [2]. Dr. Elkan then published a reply entitled "The paradoxical controversy over fuzzy Logic." [3] Elkan's basic point was that the fuzzy logic approach has been successfully used in many practical fields across the world and that almost all successful applications are embedded in the controller. However, the vast majority of theoretical articles on fuzzy method are knowledge representation and reasoning based on fuzzy logic, the basis of which has been repeatedly questioned. Viewed together, there is no transition between fuzzy theory and fuzzy application, as a result a gap or fault is formed, leading to the fact that the deterministic calculations embedded in the controller and fuzzy reasoning in logic do not necessarily converge. Long debates showed that while the fuzzy controller made some achievements, fuzzy set theory itself is yet to be improved; the combination of theory and practice needs to be more convincing.

For 50 years fuzzy logic has provided a new methodology that is different from the traditional modeling method and a new way to quantify qualitative knowledge. The combination of fuzzy logic with neural networks, probability and statistics, optimization theory, and other disciplines greatly enhances the ability of humans to process knowledge representation and information. Fuzzy set theory has also driven cutting-edge research into uncertainty cognition, such as rough set, interval set, granular computing, and so on. Collisions between different academic ideas have facilitated academic prosperity, and promoted social progress. However, these collisions are also a reflection on fuzzy logic itself and a result of the limitation of its use to complex engineering practice. We can even say that many AI researchers have always doubted the significance of fuzzy logic and its potential to solve some fundamental issues, such as knowledge representation, rule-based reasoning, and self-learning. Such discussion and debate will not shake the objectivity, progress, and rationality of Zadeh's cognitive thinking and methods to promote the development of cognitive science over the past half-century. Instead, they will benefit the development of AI with uncertainty.

Cloud model is a cognitive model for qualitative and quantitative transformation, providing a quantitative explanation of the qualitative concepts formed during human cognition of the objective world. It uses the second-order Gaussian distribution method to reflect the randomness of qualitative concepts; meanwhile, it obtains the "this and that" of the qualitative concept by calculation, that is fuzziness, and with such fuzziness nature also has uncertainty. This is both an explanation for fuzziness and a departure from the fuzzy set. In terms of the cloud model, the root cause of the gap between fuzzy set theory and fuzzy controller lies in that fuzzy sets can only describe first-order or low-order vagueness, that is, its description of fuzziness is incomplete.

Concept is formed first because the brain and sensory organs directly feel the physical, geometrical, and other characteristics of the object. It is a sensation and a perception of the objective world. First, people have a feeling or perception, then they gradually form an image or representation that is a reproduction of the objective things in the brain. It can be either intuitive, changeable, memorable, or preferential attachment, which manifests in human intelligence and is an important part of cognition. People then express themselves through language, words, and symbols as a result of continuous abstraction and communication. They even go beyond time and space to form concepts and expressible thinking. Based on these efforts, people achieve more abstractive cognition, that is, knowledge. During such a thinking process, uncertainty, including fuzziness and randomness, pervades the entire thinking activity and evolutionary process.

Concept is formed also because that same person has different cognition for the same kind of objects in different phenomena, or different people have different cognition for the same kind of objects. Differences do exist. Different intelligence will enable different feeling and perception, thus leading to large differences in representation and imagery. When people think, sometimes they will find it difficult to express or to say in words, but they can understand each other in such situations. In expressible thinking, in some cases, the nature of a variety of factors and objects, and the relationship between different properties may be precipitated. Language, words, and symbols with concepts can be applied to express the connotation and denotation of concepts, thus forming conceptual thinking. Therefore, the concept expressed by natural language inevitably has uncertainty, which is both the objective attribute of human cognition, and the charm of language.

If we use mathematical language to describe the two major causes for the formation of these human concepts, the closest is the famous Central Limit Theorem in statistics, which states that if a large number of independent causal factors decide the random variable value, and the effects of each causal factor are relatively and evenly small, then the random sum approximates to a Gaussian distribution. In a specific context and theme, the consensus reached on a concept cognition will be common sense, which reflects the degree of approximation to obey Gaussian distribution; the poorer the consensus degree of cognition, the greater the hyper entropy, or the order of entropy will be higher. In other words, the farther it will be from the Gaussian distribution. The statistical cause for formation of concept is the basis for construction of the cloud model.

As humans understood and transformed nature and society, people first tried to find the simplest deterministic judgment—the regularity of "yes or no"—and set up the corresponding mathematics. Cantor's set theory laid the foundation for modern classic mathematics that deal with deterministic phenomena, and has made brilliant achievements. If people require mathematics to handle the more complex phenomenon of uncertainty, especially that in humanities, social sciences, and thinking science, or if people require a computer to identify and deal with uncertainty issues in the objective world, like a human brain, it is necessary to use a computing method to model human cognition in the concept formation process.

Therefore, to study intelligence, and human intelligence that uses natural language to express thought, an indispensable part is to formalize the concept's certainty and uncertainty, expressed by a linguistic value, which can be fulfilled by the Gaussian cloud and high-order cloud models.

7.1.2 Dependence of Fuzziness on Randomness

Fuzzy set theory and the corresponding mathematical methods established by Zadeh use accurate membership function to represent the uncertainty concept. Once the membership function is selected and "hardened" into an exact value, all uncertainties about the concept will disappear. Thereafter, there will be no fuzziness in the definition of the concept, the rule base composed of the relationship between concepts, qualitative reasoning based on the rule base, and other thinking processes that simulate human uncertainty. This kind of fuzzy logic, which does not consider the fuzziness of the membership degree itself, can only be called a type-1 fuzzy set.

To overcome these limitations, J.M. Mendel, who proposed the second-order set theory [4], opined that membership is an interval. He focused on studying interval type-2 fuzzy sets and advocated the conversion of interval uncertainty into footprint of uncertainty (FOU). FOU is selected to build a semantic uncertain model, and FOU selection is not unique. In practice, this concept is used in many cases where words are used to express rules, or the rules are pre-given by experts. If the word modeling approach is applied to express fuzzy sets, we can use type-2 fuzzy sets whose forward problem refers to quantitative data generated according to given model parameters—for example, the types and parameters of type-2 fuzzy set Upper Membership Function (UMF) and Lower Membership Function (LMF), such as trapezoidal, triangular, bell-shaped—then generate type-2 fuzzy sets model, analyze the geometric properties of their FOU, and describe their uncertainty. The inversion problem of type-2 fuzzy sets refers to parameters generated according to sample data. To be specific, there are two methods, the interval end-points approach, and the person membership function approach. The forward problem is a mapping from qualitative to quantitative, and the reverse problem is a conversion from quantitative to qualitative.

Type-2 fuzzy logic systems include fuzziness, rule base, reasoning engine, and output processor, using the mathematical method based on type-1 fuzzy logic reasoning systems to derive the output of a trigger rule for interval type-2 fuzzy logic systems. The precondition and postcondition of rules are broken into embedded type-1 fuzzy sets. The precondition and postcondition composed of these type-1 fuzzy sets use type-1 fuzzy logic system reasoning. All the triggered reasoning results of the embedded type-1 rule are integrated into one set to constitute the trigger type-2 rule set, or only use the upper and lower membership functions of the precondition and postcondition of the trigger rules to do the reasoning computing.

The team led by Mendel published hundreds of articles about type-2 fuzzy sets and systems, interval type-2 fuzzy sets used in intelligent control, computing with words, communication signal analysis, and so on. The articles have not only been recognized by their peers but have also won the 2002 IEEE Outstanding Paper

Award, the pioneer award in 2006 IEEE granular computing meeting, and the Pioneer Award of 2008 IEEE Computational Intelligence Society.

Interval type-2 fuzzy sets use the concept of FOU, upper membership function curve, and lower membership function curve to limit the scope of FOU (Figure 7.1a). For a deterministic sample, its membership in the fuzzy set is a continuous interval between [0, 1] where its membership is the same. The biggest difference between Cloud models and type-2 fuzzy sets is that the former characterizes the qualitative concept with the entirety composed of cloud drops (sample set) subject to certain distribution, so there is no clear boundary to the uncertainty area (Figure 7.1b). For a deterministic cloud drop (sample), its membership is still uncertain, but can be obtained by calculating the probability distribution, which not only does not need to be given by a human, but also has different membership. However, the overall shape of the cloud drops and the membership function of type-2 fuzzy sets have a striking similarity.

If we recall the paradoxical controversy over fuzzy logic in history and analyze the academic thoughts of people like Zadeh and Mendel, it should be admitted that the cloud model has deviated from the track of new imprecise mathematics—fuzzy mathematics that Zadeh and Mendel had tried to invent—instead it researches the uncertainty of concept, which serves as the basic unit of human cognition from the perspective of cognition science.

The concept is often represented by extension and connotation. Extension represents the concrete and tangible things that belong to such concepts; connotation represents the essence of the concept, that is, its abstractive properties. The uncertainty of concept includes the uncertainty of both the extension and the connotation. Cloud models directly use the probability and mathematical statistics method, and Gaussian probability density distribution function to form the cloud drops, deviating from Gaussian distribution by constructing a second-order, or high-order cloud generator and applying the conditional probability. Cloud model pointed

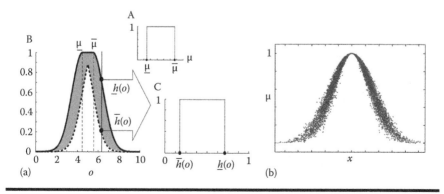

Figure 7.1 Comparison between representation methods of (a) type-2 fuzzy sets, and (b) cloud model.

out that in uncertain thinking activities, might not be a deterministic and precise membership, or membership functions, but it is easy to make people try their best to use precise mathematics to process fuzzy phenomenon. As a result, the high-order fuzzy nature of things might be ignored. In the cloud model method, especially in the forward cloud generator realized by the hierarchical use of Gaussian random values, the overall model composed of the generated discrete cloud drops is used to represent a qualitative concept. The process for forward cloud generators to generate cloud drops is a nonanalytic form of problem solving that is realized by computer programs, but cannot be clearly expressed with mathematical equations. Therefore, cloud models should not be simply viewed as random + fuzzy, fuzzy + random or secondary fuzzy, and so on. Cloud model, as a new cognitive model, represents a qualitative concept with expectations, entropy, and hyper entropy. Expectation is the center of all cloud drops (sample) distributed with uncertainty in the domain space, reflecting the basic certainty of concept; entropy is a measurement of the uncertainty of a qualitative concept, capable of reflecting both the discrete degree and the range of cloud drops (sample) acceptable to the concept, that is fuzziness, which reflects both the dependence of fuzziness on randomness and the granular of concept. Hyper entropy is the entropy of entropy and the higher-order uncertainty, a measurement of the degree of consensus reached on the concept. Of course, we can also use much higher order entropy to characterize the uncertainty of a concept, which theoretically can be infinitely higher.

People are thinking with language and involving less mathematical computing. Usually the three digital characteristics are enough to reflect the uncertainty degree of concepts. The idea that "the greatest truth is the simplest" shows human intelligence, and excessive digital characteristics will go against the nature of human thinking realized by the use of natural language. Gaussian cloud has a universal property and the universality Gaussian distribution and Gaussian membership function co-laid the foundation of the universality of the Gaussian cloud model. No matter what specific semantics a concept has, we have demonstrated that the digital characteristics of a concept has nothing to do with its semantic attributes, namely expectations, entropy, and hyper entropy consist of the universality of concept, thus realizing concept formalization, which proves the universality of the cloud model as a human cognitive model. Currently, the cloud model has made progress in both theory and application. It has been used successfully to solve the global problem of the inverted pendulum, and, as a result, won the Outstanding Paper Award at *the World Congress of the International Federation of Automatic Control*. It also played a significant role in simulating driver behavior during the self-learning process of a wheeled robot.

7.1.3 From Fuzzy to Uncertainty Reasoning

There is a question about fuzzy reasoning in the expert systems that have been developed, where very few really use fuzzy logic as a formalized tool for uncertainty reasoning. The expert system herein specially refers to the execution of complex reasoning tasks from a knowledge base of a certain scale, such as diagnosis, ordering, or data mining. In some developed expert system, various "fuzzy operators" were created to meet special needs, which is difficult to be recognized by academic peers. The design of an embedded fuzzy control chip cannot be called a complex expert system, because it does not involve many uncertainties of human cognition. Usually, a fuzzy controller is actually a deterministic calculator. As long as the appropriate smoothness constraint is given, many mathematical functions and formulas can be used as universality approximators of M-dimensional inputs and N-dimensional mapping, just like any continuous function capable of being approximated with any precision by an arbitrary high-order polynomial.

This questioning actually refers to the same things as mentioned above—that there is no transition between fuzzy theory and fuzzy application so, as a result, a gap or fault is formed, leading to the fact that the deterministic calculations embedded in the controller and fuzzy reasoning in logic do not necessarily converge. Cloud model and cloud reasoning can give a clearer answer to this question.

For a complex system with M-dimensional input and N-dimensional output, if there is a complex relationship between the variables, it presents as nonlinear, has many uncertainties, is unable to obtain a transfer function or function set between input and output with conventional analytical methods, and the data entry can be filled or exhausted that is required by the status space. So, how to achieve the uncertainty reasoning between input and output?

In uncertainty reasoning, we use a cloud model to represent the concept, and the rule to show the relationship between concepts, namely, the qualitative knowledge. When cloud reasoning and cloud control are conducted, it requires neither a precise mathematical model to be given for the controlled object, nor subjective membership, or any extended fuzzy operator. The concepts in the precondition and postcondition of the cloud rule generator may all contain uncertainty. When you enter a specific condition to activate multiple qualitative rules, by the reasoning engines, you can achieve reasoning and control with uncertainty. Output control values obtained each time will have uncertainty, which can be seen as the output of a sample near expectation (cloud drops), in terms of cognition and this might reflect the nature of the uncertainty reasoning.

For a complex system with M-dimensional input and N-dimensional output, humans have effectively summed up empirical knowledge represented with concepts in their long practice. People express a variety of concepts and relationships between them with rules, and form the rule base. Although the sample data corresponding to the concepts and rules are far from filling many data entries in status space, which reflects the sparsity, if the contents of the rule base are objective and complete enough to reflect the nature of complex systems, these rules can reflect the important turning points of the nonlinear relationship, shown by the variable relation of the complex system. The large amount of default data space outside the important turning point data, in essence, can be seen as a virtual concept or a virtual rule generated by the linear fitting of adjacent rules. This is the basic method for cloud control, for example, in an inverted pendulum, in cloud reasoning when more rules are activated, where the linear control becomes the default and the nonlinear is emphasized.

As a result, the method of cloud reasoning exactly explains the paradoxical controversy. For the uncertainty reasoning and control corresponding to the same complex system, the data in an embedded controller that has been successfully applied is just the random realization generated by cloud reasoning, which exactly reflects uncertainty resolved by cloud reasoning solutions (i.e., diversity), as long as these data do not deviate too far from the corresponding concepts and rules.

Sparsity can be found everywhere, in both nature and many complex systems. Cloud reasoning and cloud control view the valuable qualitative experiences of humans on the control of complex systems as a turning point for nonlinear control and highlight them by quantifying them through the cognitive cloud model.

7.2 From Turing Computing to Cloud Computing

If the emergence of computer provides a deterministic physical entity for the realization of AI, the Internet provides the research of AI with uncertainty as a more general and ubiquitous carrier, where people can interact at any time as individuals. The emergence and development of the Internet is full of uncertainty, which has neither a clear top-level design, nor an artificially regulated beginning, middle, and end. Throughout its development history, its architecture has been regulated through protocols. During the development and evolution of the Internet, uncertainties in network size and structure, information processing, service delivery, and public behaviors have been generated.

In 1969, the ARPA network appeared for the first time, which enabled data to be transmitted between computers. From 1984 the IP protocol started to be widely used on the Internet. Everything is over IP. Through packet switching, it provides a non-connection-oriented communication to deliver best-effort service, which is a breakthrough over the connected circuit based switched communications. There are uncertainties in network size, topology structure, nodes access, and data packet transmission paths. In 1989, the world wide web emerged, which easily realized information publishing and sharing through web technologies and represented information with semi-structured data. There are uncertainties in the size and distribution of website content, the structure of hyperlinks, and so on. In the twenty-first century, with the advent and development of web services, the Semantic Web, Web2.0, cloud computing, and other semi-structured data, especially the influx of big data in an unstructured form represented by streamed media, a social computing environment featured by Internet-based public participation gradually came into being. The Internet has developed into a social platform for idea exchange and clustering. People's perceptive and cognitive capabilities have broken away from time differences and the shackles of physical distance on earth. The uncertainty in human intelligence is reflected in the cognitive uncertainty of individuals, in niche communities, and by the public on the Internet. In September 2009, the Network Science and Engineering Committee, under the National Science Foundation of the United States, released a report entitled "Network Science and Engineering Research Agenda," which stated that over the past 40 years, computer networks, especially the Internet, have not only changed the way people live, produce, work, and entertain, but have also changed our ideas about politics, education, medical health, business, and so on. The Internet has manifested strong technological and application values, serving as a powerful engine for technological innovation and social development.

Jim Gray, a Turing Award winner, thinks that "the amount of data generated every 18 months in the network environment is equal to the total amount of data in the past thousands of years." Massive data, or big data, is often mentioned in the Internet age, such as big data storage, big data mining, big data computation, and so on. However, computing tasks or information searching often face specific situations and themes. In human cognition and reasoning activities, we need to obtain a satisfactory solution in accordance with existing concepts and knowledge, in response to an actual situation and with unstructured big data, or large enough semi-structured data, or a certain amount of structured data. It is almost impossible to achieve a uniquely optimal solution from fresh, noise-accompanied, even contradictory, original, and unstructured big data. A satisfactory solution makes it easier to serve people's expectations. Take search engines as an example, after the web crawlers process semi-structured data from the whole Internet, and then arrange them in order, the search engine interface will bring people to the ocean of results. People have to seek results in the "Next" page, but may not find a suitable one and become impatient. It is hoped that through data mining, and services for specific situations and themes, and context or semantics that consider the needs of users, we can find the user's individual preferences from social behavior, social tagging, and other information, to provide users with search services with variable granularity, best-effort, and close to personality preference. The cluster server and virtualized cluster server technology represented by Map-Reduce can automatically allocate and manage computing resources to achieve "flexible" scaling and optimized use through dynamic load balancing and resource allocation mechanism, according to demand changes.

It uses a divide-and-conquer strategy to solve complex problems. First post-split parallel computing and then integration and simplification of the results showcase good scalability, fault tolerance, timeliness, and massively parallel processing, so it is widely used in data management and analysis. The new generation of search engines will be able to do intelligent computing for each new search task based on search history and the preferences of the registered users, and present results that are close to user demand, personalized and with cross-granularity, with a delay of only seconds.

Based on standard protocols such as IP, HTTP, and web services, applications on the Internet can shield their differences in underlying transmissions and exchanges. There are diversified types of application and people pay more attention to interaction on content level, or even direct use of sound, images, pictures, written words, and expression to communicate directly with machines or robots. This kind of natural language-based interaction has become soft computing, or computing with words. Under certain situations, according to the context and grammar, syntactic and semantic understanding is realized, so various kinds of network robots emerge who can understand people's intentions. Uncertainty calculations—such as qualitative and quantitative cognitive switches between data and concepts generated by interactions through natural language and knowledge—and soft and variable granular computing will become the core issues of social computing in the Internet environment.

7.2.1 Cloud Computing beyond the Turing Machine

Turing published a famous paper in 1936 entitled "On Computable Numbers, with an Application to the Entscheidungs problem" [5], which noted that some mathematical problems are unsolvable and answered the logical completeness question in the 23 mathematical problems proposed by Hilbert in 1900. A theoretical model of an automatic computer (later widely known as the Turing machine) was first presented in this paper. The abstract model of a Turing machine can turn reasoning into a series of simple mechanical actions. There are many equivalent descriptions, such as recursive function and the abacus. As shown in Figure 7.2, the Turing machine is composed of a tape extendable indefinitely in two directions and dividable into squares (containing written symbols), a finite state controller, and a read–write magnetic head. The action of a Turing machine is determined by the quintuple $\langle q, b, a, m, q' \rangle$, wherein, q and q' are the current state and the next state of the controller; b and a are the original and the modified symbols in the squares; m indicates the moving direction of the magnetic head, to left or right or stop. The work process determined by the state and symbols is called the Turing machine program, which requires no human intervention or guidance during its automatic work. In fact, there are many statements about the Turing model. People proposed a Turing thesis, "based on the Turing machine," stating that all computable functions can be calculated by a Turing machine. From the 1960s the computer science community gradually used Turing machines to describe universal calculation abilities, and then over-treated it as a common model to solve all computing problems.

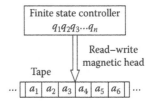

Figure 7.2 Turing machine model.

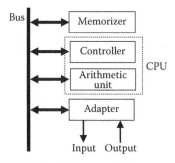

Figure 7.3   Von Neumann structure.

The Von Neumann architecture later developed, based on the Turing machine model, is a physical realization of a Turing machine model as shown in Figure 7.3. It includes controller, arithmetic unit, memorizer, input and output devices, and operates in accordance with the principle of a program stored in the form of data and executed by address order. The controller extracts the instructions and data from memory and executes them in order one by one, automatically completing the processing work described by the program. Thus, the core of the computer includes a central processing unit composed of an operational unit and a controller (i.e., CPU). People call the CPU and the operating system the computer's "core," which composes the entire computer system, together with multiple levels of hardware and software, such as micro programs, general machine, operating systems, assembly language, and senior language levels. The Turing machine and the Von Neumann structure laid the foundation for modern computers.

AI has developed many formalization reasoning methods based on the Turing machine model to solve specific problems by simulating mechanical steps taken by people when they solve deterministic problem. Later a series of methods for learning to mimic human behavior were progressively proposed, focusing on manifestations of human intelligence with a Turing machine, called machine learning. For the past 60 years, AI based on Turing machine intelligence has made great progress, and people think that the computer has replaced, extended, and enhanced human intelligence to some extent.

When faced with a lot of human cognition problems, Turing machine intelligence encounters insurmountable problems. For example, some very simple tasks for people, such as image and text recognition, are difficult to turn into a problem accurately calculable by a Turing machine. Similarly, the initiative intelligent behavior and image thinking abilities—such as commenting, writing, composing music, chatting, and programming—that are featured by human cognition, are very difficult to simulate with a Turing machine.

In fact, people do not understand the limitations of Turing machines for the following three reasons. First, the Turing machine solves input and output from the initial to the final character string, which can be considered as point-to-point mapping in the definitive space, but cannot solve the uncertainty problem. Second, the Turing model represents the computing contents with discrete numerical values, the clock frequency determines the moving speed of read–write magnetic heads, and subdivisions of same granularity determine the digitization degree of discreteness. But the Turing model does not consider if this approximation is feasible, how the numerical value returns to the analog amount in practical problems. Third, there is no formalization method for input and output in a Turing model, so there is neither man–machine interaction nor machine–machine interaction in the calculation process. AI scientists try to formalize human intelligence. Both expert systems and neural networks try their best to express formalized human intelligence in the Turing machine with algorithms, language, or symbols it can understand, and then fulfill the execution to achieve the performance of human intelligence and solve various problems. In this sense, AI scientists become "slaves" bound by a Turing machine. Even Turing himself did not think that a Turing machine was a general model suitable for solving computational problems. In 1947, he said that if you expect a machine never to make mistakes, then it must not be smart [6]. Scholars, including Turing and Turing Award winner Robin Milner, have tried to expand the Turing machine model with other interactive devices to form super-Turing computing [7]. Unfortunately, super-Turing computing research failed to foresee the invention of the Internet. In fact, the rapid development of the Internet has changed the entire computing landscape. Man and machine coexist, so both the network architecture and public behavior carry a lot of uncertainties, which cannot be described by the Turing machine.

The "Gilder speed," used to describe the growth rate of backbone bandwidth on Internet, often doubles every 6 months, while the "Moore speed," which describes CPU processing power and development of microelectronics integration density, doubles every 18 months. Since transmission bandwidth speed grows much faster than the information processing capacity, according to the evolution rule of nature's preference of attachment, we can use the transmission bandwidth in exchange for numerical computation and information processing capacity. Therefore, the development of information technology is bound to be more dependent on the network. The fact that everyone is networked makes the Earth a real global village and geographical distance is no longer important. Let us recall the development process of the IT industry. In terms of the computing environment and facilities, mainframes in the 1960s, small machines in the 1970s, PCs and LANs in the 1980s, the desktop Internet in the 1990s, have all had a profound impact on human production and life; and there is the transformation process of the mobile Internet with which people are highly concerned today. All these tell us that computing facilities and computing environment have evolved from computer-centered, to network-centered, to human-centered. In terms of software engineering, the long host-oriented software development directed at machine, language, and middle wares is being transformed to open crowd-sourcing, oriented to network, demand, and service, thus truly realizing Software as a Service (SaaS). In terms of human–computer interaction, initially it was mainly keyboard interaction; in 1964 the invention of the mouse changed the style of human–computer interaction, so that the computer was popularized, for which the inventor of the mouse won the Turing Award. Now the interaction has evolved into touch, voice, and gestures, or shake, scan, and photograph. If the traditional keyboard, Microsoft Windows systems, Apple iPhone's multi-touch operation mark the 1.0 time, 2.0 time and 3.0 time of human–computer interactions, respectively, current direct voice or text interaction technology is the driving force in human–computer interactions to enter 4.0 time. The interactive manner has gone far beyond the scope of the previous Turing machine, and has transformed from people serving computers into computers serving people.

These changes in the computing environment and facilities, the development of software engineering, and changes in interactive modes tell us that we have now entered a new era—the era of cloud computing.

Cloud computing is an Internet-based, public participatory computing model. The reason why it is called cloud is not just because of the phenomenon people usually understand as the "cloud in the sky, people on the ground," more fundamentally, it is because cloud computing has the same uncertainty as clouds. Its computing resources, including computing, storage, and interactive capacities, are dynamic, scalable, virtualized, and can be served. These are typical examples of uncertainty, and amazingly similar to the clouds in sky, composed of enormous cloud drops and showing swirling and elegant shapes.

The virtualization of computing resources has achieved rapid development in the Internet era, which enables people not to care about the following: the service mode of specific software, namely, software as a service (SaaS); the physical configuration and management of underlying resources, such as the operating system of the computing platform and the software environment, that is, platform as a service (PaaS); the location of the computing center, that is, the infrastructure as a service (IaaS). Like the industrialization revolution promoting the large-scale production of a traditional manufacturing industry to transit to social, intensive, and professional production, cloud computing is making information technology and information services achieve social, intensive, and specialized transitions so as to make information services a public infrastructure for the whole of society.

In recent years, with the universal application of 2G, 3G, and 4G wireless communication technologies, the mobile Internet achieved rapid development, while information services also developed from voice services and simple information businesses to colorful Internet applications, such as instant messaging, video communication, location-based services, social networking, and so on. End devices that people use also developed to include various intelligent end machines that focus on user experience, bringing the information industry into a new field. Cloud computing centers will become the platform for the release and management of mobile Internet applications, providing all kinds of professional and sophisticated services for computing, storage, information searching, and to derive more value-added.

In the beginning the Internet was equipped with a simple core and rich ends, supporting best-effort services. Cloud computing now allows a wide variety of cloud computing centers to offer diverse and multi-granularity information services, so the network has become rich, the ends have become simple, and interactions smarter. People rely on the Internet, especially the mobile Internet, to access personalized services anywhere and anytime, through lightweight intelligent end machines. People buy computing but not computers, storage but not memory, bandwidth but not switches. On the same end device, people can enjoy different services on the network, and different end devices can enjoy the same kind of service on the network. The end equipment delivers personalized service through customized service. Different end devices can also enjoy cluster services.

Following the appearance of computers in the 1950s and the emergence of the Internet in the 1990s, cloud computing went through a period of environment, technology, and application accumulation and has become a new milestone in the field of information. Mobile Internet is real-time, interactive, low-cost, personalized, and location aware. As a result, service demands from mobile users is growing quickly. Location-based services on the mobile Internet, and other location-based derived services, will be provided more rapidly, becoming the most down to earth cloud computing.

7.2.2 Cloud Computing and Cloud Model

After the term cloud was proposed as synonymous for service computing, such as Internet-based virtual computing, soft computing, varied granular computing, personalized computing, and so on, cloud computing has been globally recognized. The biggest characteristic of the cloud is uncertainty, which indicates that the most fundamental problem of cloud computing is to deal with uncertainty. However, up to now, not many people have understood clearly the meaning of uncertainty in the cloud and cloud drops.

Cloud model, as a cognitive model for qualitative and quantitative transformation, is mathematically based on probability and statistics. The cloud corresponds to qualitative concepts in human cognition. With cloud drops, as the random quantitative realizations of qualitative concept, forming extensions. Bidirectional transformation can be achieved by forward and backward cloud algorithms, or multiple concepts of different granularities can be extracted through cloud transformation, which laid the foundation for natural and friendly human interactions through the Internet, and completion of cloud calculations featured by "computing as a service," as well as the Internet of Things featured by materials and energy regulation through information and information technology.

People are used to using natural language to describe and analyze the objective world, particularly complex processes in nature, society, politics, economy, and management. In the process of researching and exploring problems, people tend not to care about the absolute value of quantity, but about qualitative concepts and characteristics. Therefore, use of a qualitative concept to replace a quantitative numerical value is the key technology to be solved for cloud computing to serve humanity. Qualitative descriptions must be uncertain in nature. Computing with words and the soft computing proposed by Zadeh aimed to establish a theoretical basis for future intelligent computing and calculation by word-based information systems. Zadeh opined that "mankind will eventually realize that the computing with words by use of natural language is not an optional choice, but a necessary one." People must use natural language and qualitative methods to establish the knowledge and information systems based on computing with words (concept), which is an important research direction for a cloud computing service center. In a cloud computing center with a wide variety of specific contexts and topics, the cloud model, as the qualitative and quantitative transformer, is as important and extensive as an A/D and D/A converter in a physical system. They differ in that the former is at a cognitive level, while the latter is at a digitization level, and the former is an uncertainty processing method, not a certainty processing method.

Cognitive scientists and brain scientists believe that the objective world involves physical objects, while the subjective world starts from the cognition unit and the object it points to, reflecting the characteristics of internal and external links between the subjective and objective worlds. Any thinking activities are pointing to a certain object, which is a process from the existence of an object to that of the subjective consciousness, so people's brains have been compared to a small universe. Undoubtedly, thinking in the human brain is not pure mathematics, natural language is the carrier of thinking, oral and written languages are products of human evolution. The representation of qualitative knowledge by people is, in essence, the interaction between people and their environment, including all human movement, perceptions and other activities, which will be reflected in the neural activity of the human brain. The human brain abstracts objective shapes, colors, sequences, and subjective emotional states, and generates higher-level representations through summarizing results, which is the concept, and manifests it through linguistic values. Linguistic values are the basic units of natural language and the basic "cells" in human thinking. According to the linguistic value, the human brain further selects concepts and stimulates them to produce corresponding words, or vice versa, abstracts corresponding concepts according to the words received from others.

Mankind has created languages. The process of language use is the process of using symbols, like language, for thinking. From the viewpoint of human progress, concept-based languages, theories, and models are methods for humans to describe and understand the world. Concepts are the advanced products of human brains, the objects reflected in the human brain. Concepts may be subjective, but according to statistical results, the uncertainty of certainty is a certain rule that reflects the general and most important attribute of things in essence. People use concepts to abstract, classify, or cluster all kinds of objects in the real world, according to their natures; and to construct cognition, whose complexity can be measured by psychological scales, so that cognition can rise from a low-level perceptual stage to a high-level rational stage. Such a process of concept formation in the human brain is the most basic performance of thinking.

Linguistic values directly associated with concepts can concentrate cognition and reduce the complexity of a concept structure to a controllable extent. Thus, humans need to extract and find qualitative knowledge from a large amount of complex data that can be expressed in natural language. Qualitative knowledge has the natural property of uncertainty, including uncertainty, incompleteness (unfullness), uncoordinatedness, and impermanence. Cloud computing will play a bigger role in dealing with these uncertainties.

7.2.3 Cloud Model Walking between Gaussian and Power Law Distribution

Gaussian distribution is widespread in nature. If independent causal factors can decide the random variable value, and the independent role of each causal factor contributes to the entirety in a relatively small and even way, then the random variable is approximately Gaussian.

In nature, there is another distribution, namely, power law distribution, which refers to the power function relationship between the probability density functions and variables. One of the most important features of power law distribution is that it presents as linear in the log–log coordinate, which indicates it has both heavy-tailed features and scale invariance, that is, it not only decays slowly in the tail, but also decays in the same proportion in different scales, so, in complex network theory, it is also known as scale-free distribution; it was also referred to as the 20–80 rule, used to explain the feature of fat tail. The power law distribution was first discovered in the field of linguistics, called Zipf's law, that is, within a specific pragmatic range, if you use a sufficiently large amount of vocabulary, arrange it in order from high to low according to the utilization rate, then the relationship between the probability of word emergence and its corresponding order number presents a power law distribution. In recent years, a series of articles have been published in *Nature* and *Science* revealing that most complex networks in the real world have a scale-free property, which aroused interest in power law distribution from different fields, such as mathematics, statistics, physics, computer science, ecology, life sciences, and so on. As a result, the position of power law distribution has been raised to be as important as that of Gaussian distribution.

Since the cloud model was proposed, it has attracted a lot of attention and realized many applications. Let us review the Gaussian cloud model as defined from the perspective of an algorithm:

Input: The three digital characteristics representing the qualitative concept C (*Ex*, *En*, *He*), cloud drop number N.

Output: Quantitative value of N cloud drops, and degree of certainty of each cloud drop representing concept C.

Steps:

1. Generate Gaussian random number En' with En as expectation, He as standard deviation.
2. Generate Gaussian random number x with Ex as expectation, abs (En) as standard deviation.
3. Let x be specific quantitative value of qualitative concept C, called cloud drops.
4. Calculate $y = e^{-((x-Ex)^2/2En'^2)}$, let y be certainty degree of x to represent the qualitative concept C.
5. Repeat steps (1) to (4), until N cloud drops are generated.

The cloud drops in Gaussian clouds are produced through the second-order Gaussian process. There is a difference between the distribution of cloud drops X and the standard Gaussian distribution ($He = 0$). The four sub-figures in Figure 7.4 show that if $Ex = 0$, $En = 1$, 10,000 cloud drops will be generated under different hyper entropy conditions. Figure 7.4a is a Gaussian distribution with hyper entropy of zero. The distribution histogram shows the phenomenon of two ends small and middle part big, which is a standard Gaussian distribution; with hyper entropy increasing to be 0.1, 1, and 10, the cloud drops distribution will be formed in Figure 7.4b through d.

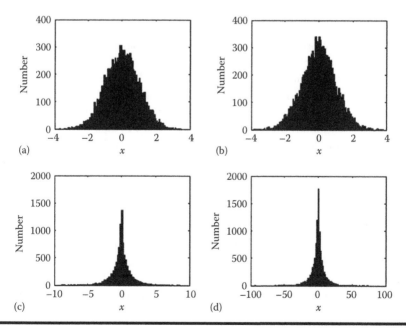

Figure 7.4 Cloud drops distribution with different hyper entropy: (a) $Ex = 0$, $En = 1$, $He = 0$, $N = 10,000$; (b) $Ex = 0$, $En = 1$, $He = 0.1$, $N = 10,000$; (c) $Ex = 0$, $En = 1$, $He = 1$, $N = 10,000$; and (d) $Ex = 0$, $En = 1$, $He = 10$, $N = 10,000$.

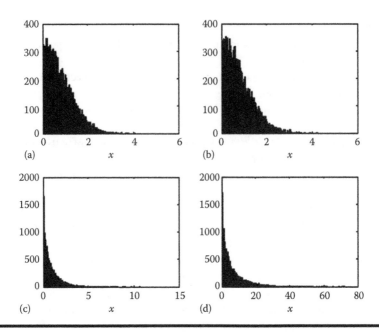

Figure 7.5 Cloud drops distribution after symmetry point decided: (a) *Ex* = 0, *En* = 1, *He* = 0, *N* = 10,000; (b) *Ex* = 0, *En* = 1, *He* = 0.1, *N* = 10,000; (c) *Ex* = 0, *En* = 1, *He* = 1, *N* = 10,000; and (d) *Ex* = 0, *En* = 1, *He* = 10, *N* = 10,000.

It can be seen that cloud drops near expectation *Ex* accounted for an increasing proportion of the overall cloud drops.

We made a slight modification to the Gaussian cloud generation algorithm. If we only focus on those cloud drops with *X* values bigger than the expectation, cloud drops less than *Ex* can be gotten by using the symmetry points with *Ex* as center, the cloud drops distribution obtained by this way is shown in Figure 7.5.

In terms of morphological shape, as shown in Figure 7.5, when hyper entropy increases, the cloud drops distribution moves closer to the power law distribution. But it is certain that the second-order Gaussian cloud will never form a power law distribution, no matter how big the value of hyper entropy is. To further explain this problem, the concept of kurtosis is introduced. Kurtosis is an important characteristic value to describe distribution status in statistics. It is used to judge the degree of the flatness and pointedness of the distribution curve compared with Gaussian distribution. If Gaussian distribution is regarded as mesokurtic with kurtosis of 0, then the distribution curve with shapes higher and thinner than that of Gaussian distribution is called leptokurtic, otherwise called platykurtic. The formula to calculate kurtosis is

$$k(X) = \frac{E\left[X - E(X)\right]^4}{\left\{E\left[X - E(X)\right]^2\right\}^2} - 3$$

The fourth-order central moment of Gaussian cloud has the meaning of kurtosis. Its kurtosis of

$$k(X) = \frac{E\left[X - E(X)\right]^4}{\left\{E\left[X - E(X)\right]^2\right\}^2} - 3$$

$$= \frac{3\left(3He^4 + 6He^2 En^2 + En^4\right)}{\left(En^2 + He^2\right)^2} - 3$$

$$= 6 - \frac{6}{\left(1 + \left(He^2 / En^2\right)\right)^2}$$

When *He* tends to big infinitely, the kurtosis of Gaussian cloud approaches 6.

From the experimental point of view, when hyper entropy changes to be very big, its *X*-projection interval frequency is shown in Figure 7.6. In Figure 7.6b, the hyper entropy is 100 times bigger than entropy, but the relationship between frequency and value under log–log coordinate still deviates obviously from the straight line (power law distribution). At this moment, if the hyper entropy continues to increase infinitely, then the cloud drops discrete unlimitedly, and Gaussian cloud will gradually deviate from power law distribution.

We introduced the concepts of hyper hyper-entropy and higher-order entropy. Gaussian cloud is the product of second-order Gaussian distribution. It uses the

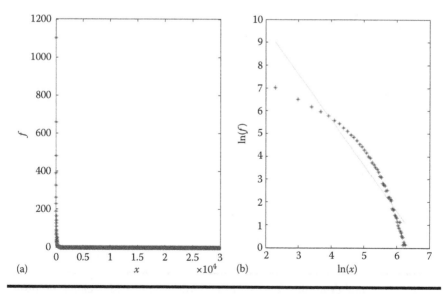

Figure 7.6 Cloud drops distribution when hyper entropy is 100 times bigger than entropy: (a) *Ex* = 0, *En* = 1, *He* = 100 and (b) log–log coordinate.

random sample values of second-order Gaussian as variance of a Gaussian. With hyper entropy increasing, its shape tends to approach power law. Considering the hyper entropy in cloud model is a measurement of the uncertainty of entropy and the variance of entropy, there can be hyper hyper-entropy (third order), fourth-order, fifth-order, or higher-order entropy. The distribution of cloud drops with third-order, fifth-order, and multi-order Gaussian is studied herein.

First we discuss the third-order Gaussian cloud: Give *Ex*, *En*, *He*, *HHe* to generate *N* cloud drops.

First, use *He* as the center value, *HHe* as variance to generate a random hyper hyper-entropy *HHe'* greater than *He* and capable of meeting Gaussian distribution (also called third-order random entropy); second, use *En* as the center value, *HHe'* as variance to generate a random hyper-entropy *En'* greater than *En* and subject to Gaussian distribution (may be called second-order random entropy); third, use *Ex* as center value, *En'* as variance to generate a random cloud drop subject to Gaussian distribution. Cycle like this till *N* cloud drops are generated. The frequency distribution of the cloud drops is shown in Figure 7.7.

Figure 7.7b shows the case of log–log coordinate, which can use the least squares to approximate the fitting, with power exponent approximating 0.887.

Similarly, high-order cloud drops, such as fifth-order, tenth-order, twentieth-order can be generated, as shown in Figures 7.8 through 7.10, respectively.

Experiment tells us that in the second-order Gaussian cloud model, with the hyper entropy increasing, the cloud drop group starts to present a distribution

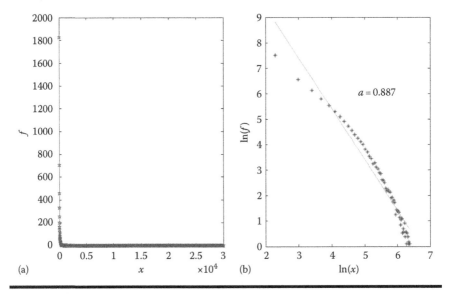

Figure 7.7 Cloud drops distribution of the third-order Gaussian cloud: (a) result of three times normal function and (b) the log–log coordinate fit (exponent *a* = 0.887).

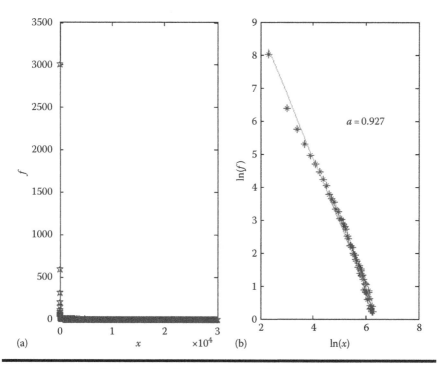

Figure 7.8 Cloud drops distribution of the fifth-order Gaussian cloud: (a) result of five times normal function and (b) the log–log coordinate fit (exponent $a = 0.927$).

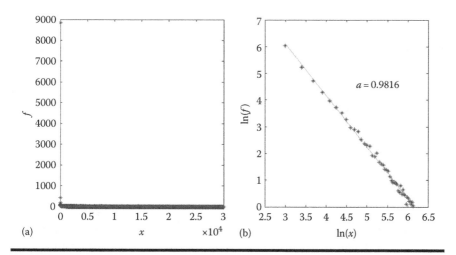

Figure 7.9 Cloud drops distribution of the tenth-order Gaussian cloud: (a) result of 10 times normal function and (b) log–log coordinate fit (exponent $a = 0.9816$).

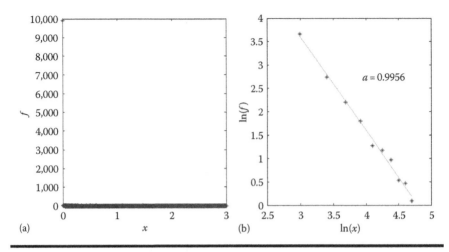

Figure 7.10   Cloud drops distribution of the twentieth-order Gaussian cloud: (a) result of 20 times normal function and (b) the log–log coordinate fit (exponent a = 0.9956).

featured by fat tail, but further increase of hyper entropy does not lead to power law distribution; the third order Gaussian can produce distribution relatively fit for the power law; higher order Gaussian continue to function, such as 5 or 10 times, which will generate shapes closer to power law distribution; with the increase of order, the cloud drops generated by a K-order cloud model will tend to show more apparent power law distribution. In theory, the probability density function of a higher order Gaussian cloud can be deduced with conditional probability, while their basic mathematical properties can be obtained, such as higher order central moments, and so on. This means that using different order iterations, a high-order cloud model can walk between Gaussian distribution and power law distribution.

7.3 Big Data Calls for AI with Uncertainties

7.3.1 From Database to Big Data

If the Turing computable model pioneered the computer age, the relational algebra model inaugurated the database era of managing structured data on computers. Since then, the computer has no longer been simply oriented to numerical or mathematical calculations in the field of scientific research, but has served the whole of society and all industries with data and information processing. Since the 1970s, the hierarchical and network database models approached the relational database model. People set up a wide variety of relational database management systems. Through data modeling tools, indexing and data organization techniques, and by means of user interfaces, query languages to optimize query processing, and transaction management, people can access data or update databases conveniently and flexibly. The capacity of online transaction processing has become an important indicator of database management systems. However, as global digitization accelerates its processes, digital contents using large frequency or big value are structured or fragmented, added to the database, which becomes structured hot data to meet the needs of repeated use. As a result databases have become increasingly big. They were called large databases in the 1980s, very large databases in the 1990s, and extremely large databases in the early 2000s. However, the changes in name cannot keep up with the growth in data. People store information gathered from multiple data sources in a generally consistent pattern and put them in a single site, called a data warehouse. People use multidimensional databases to build the data cube or hypercube model, or to maintain a data warehouse through data cleaning, data transformation, data integration, data loading, and periodical data update. The capacity of online analysis processing has become an important indicator of a data warehouse, greatly promoting the construction of the information society.

It should be said that during the development of database and data warehouse technology, people rarely considered the uncertainty of data in data use. If anyone did consider such uncertainty, it was just about fuzzy matching and fuzzy search technologies.

The rapid upsizing of databases and data warehouses will inevitably lead to the embarrassing situation of "abundant in data, poor in information, lack of knowledge." Thus, knowledge discovery from databases (KDD) technology has become a technology hotspot. Classification or clustering of data objects, exploration of the association rules or sequence law of databases, the analysis of the abnormal behavior of data objects as isolated points, or predicting the evolution of trends in data objects, led people to find the background of knowledge discovery, focus on how to describe the mining task, the types of knowledge, the concept granularity, and conceptual hierarchy, and even how to measure the interest degree of discovery.

However, the rapid rise of Internet technology, especially those related to the world wide web and the mobile Internet, promoted information sharing and interconnection based on hyperlinks, hypertext, and the emergence of virtual cluster search servers based on website crawler technology. Search engines came out as a dark horse in response to semi-structured data, attracting attention from all of society. It challenged the traditional structured database technology and research into discovery of knowledge in databases. People live in an Internet age when searching is needed anywhere, at anytime, by anybody. People are willing to pay for search results. Under the banner of cloud computing, it has become a trend to take searching as a service, even as a personalized service to access information. Traditionally, top-level design is first completed to build the framework for the base or warehouse, and then the data is pre-processed and structured, moved into a premade framework, but now such practices are being questioned. In 1995, SIGKDD convened an annual international conference on knowledge discovery and data mining. Since then, the meeting has been upsizing, with fewer and fewer articles about knowledge discovery, but more and more articles on the value of data, and massive data mining. In 2012, the 18th KDD meeting was held in Beijing, where the hotspot has turned to big data excavation.

The growth rate in data in an information age, such as natural big data, life big data, and social connection big data, is much higher than Moore speed reflecting microelectronics, computers, and computing capability. A variety of data from multiple sensors, in particular the data flow, has such a big scale that it is impossible to capture, manage, and process with software tools in conventional databases or data warehouses within a certain period of time and quickly get valuable information. Big data is usually fresh, and featured by high noise, high redundancy, low value density, incompleteness, inconsistency, and many other original features, especially the timely interaction and mutual function with people and environment during use. It is not favored to handle big data on a single computer with a strong computing model. Big data technologies have become a science in the PB (10^{15} bytes) era.

Is there an unstructured solution like the traditional relation database that can eventually be widely applied to many big data areas and quickly discover the value of data? Such new methods and measures must ensure the simplicity, easy scalability, timeliness, and interactivity of processing.

We can say that cloud computing and big data are twin brothers of the Internet age. Led by the concept of software as a service, taking searching as a service will lead to taking mining as a service. People might not urgently need too much knowledge as universal as that in the British *Encyclopedia Britannica*, but they do demand timely satisfaction of their needs through data mining in the open environment, such as finding the most interested friends, views, news, information, opportunities, trends, and so on. These are certainly not the only optimal solutions, but they can be the most satisfactory solutions with best efforts; certainly not a definitive solution, but can be the essentially definitive solution with uncertainty; they are not fit for all, but are a personalized solution. Big data promotes the great development of AI with uncertainty.

7.3.2 Network Interaction and Swarm Intelligence

With the popularization of the Internet and the development of network science, especially the growth of social networks, people are increasingly concerned about individuals equipped with subjective behavioral abilities, and group behaviors arising from communication and interaction between individuals.

In the past, people studied group behaviors in particle swarms, ant colonies, fish schools, bee swarms, and so on, paying attention to biology-related evolutionary computation and even natural computing. Now, people are more concerned about the various kinds of niche communities (associations) with different scales formed through sharing and interaction on the Internet, and the swarm intelligence that emerges there from, that is, social computing. For example, the follow relationship on micro-blogging, the friend relationship in social networks, and the common interest relationship in online stores due to co-buy or product comments, have shown the prevalent local clustering and the swarm intelligence they reflect. The public faced by the Internet is no longer simple, extensive, and have the same requirements, but is a public composed of niche groups with distinctive features, or we can say that there is no niche, there is no public.

In the past, the popularity of communication facility enabled thousands of families to have the same kind of telephone, which was described as "big river with water can make stream full." Now, under the condition of social networking available everywhere, the service to satisfy the niche group is a phenomenon of "stream with water can make the river full."

Computers are embedded into networks, the environment, or everyday tools. Computing equipment is fading out of center position in the calculation process. People have started to focus on the task itself. People have built a ubiquitous environment capable of computing, storage, and communication, via the Internet, which is gradually blended with people. For example, folksonomy uses the intelligence groups formed by public contribution of contents on their initiative and data sharing to fulfill difficult tasks that cannot be completed by Turing machine computation, such as multimedia identification, marking, and classification. It must be noted that swarm intelligence is often a statistical form emerging as a result of interaction in an open, Internet-based computing environment without centralized control. In the cloud computing environment, everyone is both a user and a developer of information resources, both a consumer and a provider of services, reflecting the synergy intelligence produced by human–computer coexistence.

Wikipedia uses the public to collectively create *Wikipedia* entries, which is another typical application for public participation to form a swarm intelligence. In *Wikipedia*'s operation mode, any individual can edit the entries that interest them. Such editing is free and individuals can contribute any viewpoint to any entry. Although the participants may make mistakes in the editing, and some even tamper maliciously, in the case of public participation, mistakes and malicious tampering will soon be corrected; most entries have maintained a very high level of quality. For example, in Figure 7.11, the entry for cloud computing was founded on September 4, 2007. Its definition evolved from the initial simple and a one-sided controversial version to the more accurate, stable, and rational explanation on February 23, 2011. Like the cognitive process for humans to explore the true meaning of a concept, after repeated sharing, interaction, inclusion, revision, and evolution, the *Wikipedia* entries co-edited by the public will become relatively more correct. Such entries will be recognized by most participants or readers, thus reaching a relatively stable consensus close to the truth.

We can say that it is the Internet without centralized control that feedbacks and incorporates the public's cognition into the calculation process, while people can select with uncertain thinking; therefore, the computing input state on the Internet is no longer represented by a deterministic set, and the production of

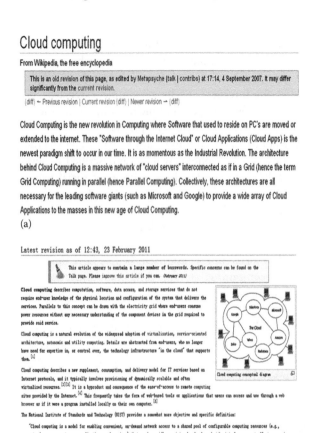

Figure 7.11   Evolution of the cloud computing entry in *Wikipedia*: (a) September 4, 2007 and (b) February 23, 2011.

swarm intelligence cannot be represented by a deterministic operation sequence like the Turing machine. Uncertainty is particularly prominent in this process. But the feature of uncertainty does not mean that we cannot explore the overall result produced by the swarm intelligence when it functions on an Internet that is seemingly random. Studying the physical characteristics of networks by means of AI with uncertainty may reach a valuable law.

The swarm intelligence established on the basis of public interaction and in compliance with important statistical characteristics is significantly different from Turing machine intelligence. Turing machine intelligence has advantages in numerical computation, storage, retrieval, and deterministic reasoning ability, but there have been no substantial breakthroughs when more uncertainty cognition is needed, such as image thinking or common sense, inspiration and sudden enlightenment. An Internet-based swarm intelligence can conduct repeated interactions, advancements, abstractions, and applications regarding the views of participants, and perceive and feedback on various models in the real world. It can use people's common sense to create advantages in uncertainty cognition, such as image thinking, association, sensibility, and sudden enlightenment.

Swarm intelligence is a kind of intelligence emerging in the process of constant interaction and communication among the public on the Internet, which can absorb and utilize certain cognitive abilities contributed by the participants, and reflect stable statistical characteristics, or emergent structures. Community clusters a group of people on the Internet with an interest in a specific topic in a particular context. The emergence of an Internet community smartly solves the problem of definition and representation of context-based common knowledge that cannot be resolved by AI. In the information age, everyone lives in the community and in multiple different communities. Overall, the intersection, coverage, and inclusion between communities, and the interaction and transition of diverse roles among community members has produced some typical internet behaviors, such as onlookers often "getting together" on the Internet, the emergence of opinion leaders, theme focus and dissipation, and differentiated commentary gradually converging to the state of consensus. From the perspective of AI scholars, an in-depth study of the characteristics and applications of swarm intelligence, improvement of the intelligence created by machine–machine and machine–man interaction on the Internet, will eventually help us to develop AI with a stronger thinking ability, more easily accepted by the human mind, and capable of harmoniously co-existing with human beings. This is also the goal when computers were first invented as thinking machines.

7.4 Prospect of AI with Uncertainty

Looking back on the history of human understanding of the objective world, the most concentrated and most prominent amazing achievement was made in physics. In terms of understanding material structures, on one hand, people explored large-scale goals, including planets and galaxies; on the other hand, people actively explored the micro world to find the smaller composition units of substance, from molecules and atoms to nuclei, then deep into neutrons, protons, and further into the quark level. Physicists relentlessly pursued a more unified theory to cover the four interaction forces, namely, gravitational force, electromagnetic force, the strong interaction force, and the weak interaction force.

Therefore, can understanding the human brain, also a small universe, draw on people's cognition of the objective world? We believe that an important direction for the development of cognition and thinking science in the twenty-first century is to extend the theory result of cognition of the objective world in modern physics to the cognition of the subjective world. This is the cognitive physical method that we are pursuing diligently, namely, cognitive physics.

It is well known that in the physical understanding of the objective world, model and field have played a key role. Drawing on the idea of the field, we introduced the interaction between particles and the field-related description method into the abstractive cognitive space. According to Cognitive Physics thinking, if we use the idea of field to formalize the representation of people's cognition and thinking processes, from data to concept, and from concept to knowledge, we can establish a cognitive field to describe the interaction between data and to visualize human cognition, memory, thinking, and other processes. Data, concepts, linguistic values, and groups in domain space are nothing but objects interacting with each other in the field space that are only different in granulation.

People often analyze and understand nature, human society, and human thinking activities with different scales. It is also an important content of Cognitive Physics to draw on granularity in physics to reflect the granularity of knowledge or the scale of concept. Human cognitive processes have different levels, such as feeling, perception, representation, concept, abstraction, and so on. The levels are related to the granularity of the object, and the level of knowledge is related to the granularity of the concept. No matter what kind of knowledge is discovered, if we elevate the original concept with lower granularity, we will find more general and more universal knowledge, which is the emerging granular computing.

An accepted feature of human cognition is the ability of people to observe and analyze the same problem from different granularities, so people will not only solve a problem in the world with the same granularity, but also jump quickly and freely from one granularity world to another. Human beings even can handle different granular worlds simultaneously. This is the strong performance of human problem-solving capability. In fact, the process of human cognition and thinking corresponds to the transformation process of concepts represented by different granularities between different scales, that is, a transition from a relatively stable state to another relatively stable state. How to formalize the transition from data to concept, and from concept to knowledge discovery in the process of human cognition and the process of progressive induction and concision of knowledge from fine-grained to coarse-grained is also a basic problem in AI research. We draw on the characteristics of multiview, multiscale, and multilevel of physical space and use the idea of state space transformation in physics to form the framework for transformation of knowledge discovery state space. Each state in the space represents a relatively stable knowledge form, and the cognitive process corresponds to the transformation from one state space to another state space. Cloud transformation and data field have become important tools for transformation of discovery state space.

From the end of the twentieth century to the mid-twenty-first century, mankind is in an era when science is both highly differentiated and highly integrated. Information science and technology boasts of the fastest growing and most enduring subject in all sciences and continues to act as the most powerful engine for human development. Moreover, it continues to be the dominant force supporting cross-discipline, technological innovation, and economic development; keeps penetrating into the whole of society in all its aspects; and promotes society to transit from a multiple pyramid structure to a flat network structure. In the three elements constituting the world, namely, substance, energy, and information, an earth that had previously been regarded as vast has become a global village, with limited substance and energy. We should finely control material and energy through information science and technology, so that people and nature can develop harmoniously, sustainably, and in a scientific way. We should tap the productivity from data and information, perceive the earth by means of the Internet, mobile Internet, Internet of people, Internet of Things and cloud computing, and make the earth a smart planet on the basis of a digital earth.

Horizontal development has spawned more inter-disciplinary and cross-disciplinary subjects and become a major source of scientific progress and technological innovation. The separated and vertical discipline divisions have transformed to crossed and collaborative scientific research, and traditional innovation has evolved into network-based group innovation featured by openness, open sources, and collaboration.

Scientific discoveries and technological inventions are often interdependent. In the process of scientific development, sometimes discovery is earlier than invention; sometimes invention is ahead of discovery. Discovery does not always come before invention. If we say that the Turing model of 1936 was metaphysical science, the Turing model materialized into a physical entity via von Neumann in 1944 was a physical technology. If we say that the relational algebraic model of the 1970s was a metaphysical science, the later relational algebra materialized into databases and data warehouses was a physical technology, then today, the cloud computing and big data mining technologies are results of invention ahead of discovery. In recent years, among the Turing Award, Shannon Prize and Nobel Prize winners, those awarded prizes for their achievements in experimental technologies occupy a large proportion. Technology can also be precursor of science. We look forward to the emergence of new scientific theories that can guide cloud computing and big data so as to embrace the scientific revolution generated by digitalization.

It needs wisdom and intelligence to appreciate the aesthetic value of great artistic works and understand great sciences. They both seek universality, profundity, and meaningfulness in pursuit of truth and beauty. This is where charm can be found in basic research and interdisciplinary research. It is a great pleasure of life to appreciate the splendid scenery unfolded by the convergence of different disciplines and the intersection of science and art, which need both poets and scientists to keep a child-like heart. We hope that this book will encourage readers to make an effort to achieve more in their life.

References

1. C. Elkan, The paradoxical success of fuzzy logic, *IEEE Expert*, 9(4), 3–8, 1994.
2. L. A. Zadeh, Responses to Elkan: Why the success of fuzzy logic is not paradoxical, *IEEE Expert*, 9(4), 43–46, 1994.
3. C. Elkan, Elkan's reply: The paradoxical controversy over fuzzy logic, *IEEE Expert*, 9(4), 47–48, 1994.
4. N. N. Karnik and J. M. Mendel, Introduction to type-2 fuzzy logic systems. In: *IEEE World Congress on IEEE International Conference on Fuzzy Systems*, Anchorage, AK, Vol. 2, pp. 915–920, 1998.
5. A. M. Turing, On computable numbers, with an application to the Entscheidungs problem, *Proceedings of the London Mathematical Society*, 2(42), 230–265, 1936; (43), 544–546.
6. A. M. Turing, Lecture to the London Mathematical Society, typescript in the Turing Archive, King's College, Cambridge, U.K., February 1947.
7. M. Robin, *Communication and Concurrency*, International Series in Computer Science, Prentice Hall, New York, 1989.

Index

A

Abstract concept, 32
Adaptive Gaussian cloud transformation (A-GCT)
  CAE academicians, age distribution of
    ArnetMiner users' ages, 99–100
    concept level, 99, 101
    digital characteristics, 97–98
    frequency curve, 97–98
    pan-tree for, 99, 101
  differential objects extraction
    color image segmentation, 111–112
    desert snake image, 111
    misclassification error, 109
  input, 96
  laser cladding image segmentation, 103,
    105–107
  output, 96
  steps, 96
Age concept tree, 82
AI, *see* Artificial intelligence
AlphaGo, 12
American Association for Artificial
    Intelligence, 16
Animats, 22
ANNs, *see* Artificial neural networks
Anti-blink technology, 212
Anti-rear-end technology, 212
Applause sound
  cloud model
    environment, 234
    individual behavior, 232–235
  computing model
    audience clapping times, 238
    audience members, 237–238
    clap intensity, 236–237
    clapping interval, 237–238
    coupling functions, 238–239
    courteous applause, 239

    data field, 235–236
  emergence, diversity analysis of
    clapping intervals *vs.* time, 244
    entropy, 242–244
    no-applause, probability of, 243
  experimental platform
    computing simulation, 242–243
    courteous applause, 241–243
    data stream file, 239–240
    initial state and audience interaction,
      239–241
    interwoven applause, 242
    spontaneously synchronized applause,
      241–243
  guided applause synchronization,
    245–248
Argonne National Laboratory, 16
ArnetMiner, 99–100
ARPA network, 258
Artificial intelligence (AI)
  definition, 1
  development of
    in ancient Egypt, 8
    artificial life, 13–14
    Dartmouth Symposium, 8–10
    identity recognition, 15
    KDD, 15–16
    knowledge engineering, 15
    machine *vs.* human match, 11–12
    pattern recognition, 14–15
    robots, research on, 16–17
    theorem proving by machine, 11
    thinking machine, 12–13
    Turing test, 10–11
  human intelligence, uncertainty of
    acceptance of existence, 5
    cognitive uncertainty, 6
    common sense, 5
    deterministic science, 2–3

Printed and bound by CPI Group (UK) Ltd, Croydon, CR0 4YY

23/10/2024

01777697-0006